Fluid Transport

There are no such things as applied sciences,
only applications of science.
Louis Pasteur (11 September 1871)

Dedicated to my wife, Anne, without whose unwavering support, none of this would have been possible.

Industrial Equipment for Chemical Engineering Set

coordinated by
Jean-Paul Duroudier

Fluid Transport

Pipes

Jean-Paul Duroudier

ISTE PRESS

ELSEVIER

First published 2016 in Great Britain and the United States by ISTE Press Ltd and Elsevier Ltd

ISTE Press Ltd
27-37 St George's Road
London SW19 4EU
UK

www.iste.co.uk

Elsevier Ltd
The Boulevard, Langford Lane
Kidlington, Oxford, OX5 1GB
UK

www.elsevier.com

British Library Cataloguing-in-Publication Data
A CIP record for this book is available from the British Library
Library of Congress Cataloging in Publication Data
A catalog record for this book is available from the Library of Congress
ISBN 978-1-78548-184-0

Contents

Preface

The observation is often made that, in creating a chemical installation, the time spent on the recipient where the reaction takes place (the reactor) accounts for no more than 5% of the total time spent on the project. This series of books deals with the remaining 95% (with the exception of oil-fired furnaces).

It is conceivable that humans will never understand all the truths of the world. What is certain, though, is that we can and indeed must understand what we and other humans have done and created, and, in particular, the tools we have designed.

Even two thousand years ago, the saying existed: "faber fit fabricando", which, loosely translated, means: "*c'est en forgeant que l'on devient forgeron*" (a popular French adage: *one becomes a smith by smithing*), or, still more freely translated into English, "practice makes perfect". The "artisan" (faber) of the 21st Century is really the engineer who devises or describes models of thought. It is precisely that which this series of books investigates, the author having long combined industrial practice and reflection about world research.

Scientific and technical research in the 20th century was characterized by a veritable explosion of results. Undeniably, some of the techniques discussed herein date back a very long way (for instance, the mixture of water and ethanol has been being distilled for over a millennium). Today, though, computers are needed to simulate the operation of the atmospheric distillation column of an oil refinery. The laws used may be simple statistical correlations but, sometimes, simple reasoning is enough to account for a phenomenon.

Since our very beginnings on this planet, humans have had to deal with the four primordial "elements" as they were known in the ancient world: earth, water, air and fire (and a fifth: aether). Today, we speak of gases, liquids, minerals and vegetables, and finally energy.

The unit operation expressing the behavior of matter are described in thirteen volumes.

It would be pointless, as popular wisdom has it, to try to "reinvent the wheel" – i.e. go through prior results. Indeed, we well know that all human reflection is based on memory, and it has been said for centuries that every generation is standing on the shoulders of the previous one.

Therefore, exploiting numerous references taken from all over the world, this series of books describes the operation, the advantages, the drawbacks and, especially, the choices needing to be made for the various pieces of equipment used in tens of elementary operations in industry. It presents simple calculations but also sophisticated logics which will help businesses avoid lengthy and costly testing and trial-and-error.

Herein, readers will find the methods needed for the understanding the machinery, even if, sometimes, we must not shy away from complicated calculations. Fortunately, engineers are trained in computer science, and highly-accurate machines are available on the market, which enables the operator or designer to, themselves, build the programs they need. Indeed, we have to be careful in using commercial programs with obscure internal logic which are not necessarily well suited to the problem at hand.

The copies of all the publications used in this book were provided by the *Institut National d'Information Scientifique et Technique* at Vandœuvre-lès-Nancy.

The books published in France can be consulted at the *Bibliothèque Nationale de France*; those from elsewhere are available at the British Library in London.

In the in-chapter bibliographies, the name of the author is specified so as to give each researcher his/her due. By consulting these works, readers may gain more in-depth knowledge about each subject if he/she so desires. In a reflection of today's multilingual world, the references to which this series points are in German, French and English.

The problems of optimization of costs have not been touched upon. However, when armed with a good knowledge of the devices' operating parameters, there is no problem with using the method of steepest descent so as to minimize the sum of the investment and operating expenditure.

1

Fluid Ejectors and Gas Ejectors

1.1. General

1.1.1. *Principle of an ejector*

Consider a convergent followed by a divergent. Between them is a neck with a constant cross-section, and the driving *fluid* goes through this convergent–divergent.

Using the Bernoulli equation, we can describe the flow of an incompressible fluid:

$$\rho\frac{V^2}{2}+P=\text{cste}$$

In other words, the more the velocity increases, the more the pressure diminishes.

At the neck, the speed is at its maximum and the pressure is at its minimum. Consequently, this location can be used for the arrival of *suction fluid* piping.

We will see that, for a simple fluid, the conservation of momentum should be employed.

Gases require:

– the conservation of energy;

– the conservation of momentum.

It is accepted that the transformations are initially adiabatic and that the subsequent temperature of the combined motor gas + suction gas is calculated by the rule of mixtures.

Before going further, the reader would benefit from an understanding of the treatment of Fluid Mechanics by Brun *et al.* [BRU 68] concerning:

– the flow of incompressible fluids;

– the flow of compressible fluids.

1.2. Liquid–liquid or gas–gas ejectors

1.2.1. *Parameters of the problem*

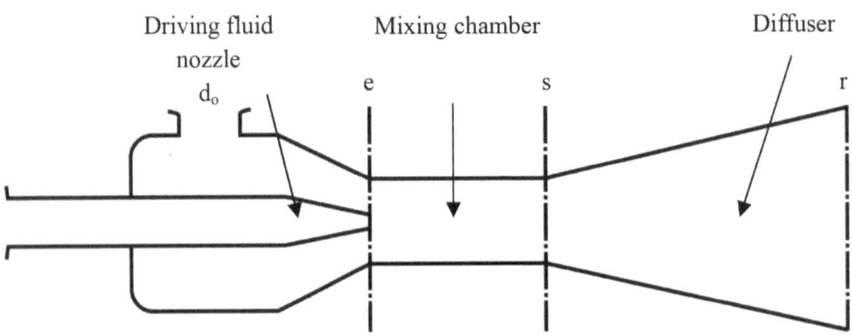

Figure 1.1. *Fluid–fluid ejector*

Here, we accept that, in the mixing chamber located between cross-sections e and s, the mixture is uniform throughout. We will call the value of this cross-section A (in m^2).

Now, we will define the various parameters required.

W_a: flow rate of suction fluid (kg.s^{-1})

τ: ratio of the cross-section of the driving fluid outflow nozzle to the chamber cross-section σ: ratio of the driving fluid flow to the flow rate of suction fluid (specific consumption)

P_{ao}: pressure generated by suction liquid (Pa)

P_{mo}: pressure generated by driving liquid (Pa)

P_e: pressure at the entrance of the mixing chamber (Pa)

P_s: pressure at the exit of the mixing chamber (Pa)

P_r: pressure on diffuser discharge (Pa)

ρ_a, ρ_m, ρ_r: measured densities of suction fluid, driving fluid and mixture discharged at diffuser exit ($kg.m^{-3}$).

1.2.2. *Finding the characteristic equation*

This equation allows us to calculate σ according to τ in view of pressures P_{ao}, P_{mo} and P_r.

The acceleration of the driving liquid is denoted by the following expression:

$$\eta_m (P_{mo} - P_e) = \frac{\rho_m U^2{}_{me}}{2} \qquad [1.1]$$

In this expression:

$$0.97 < \eta_m < 0.99$$

η_m is the yield of the operation.

U_{me} is the velocity of the driving fluid at the entrance of the mixing chamber:

$$U_{me} = \frac{\sigma W_a}{\rho_m A \tau}$$

Similarly, the acceleration of the suction fluid is denoted by:

$$\eta_a \left(P_{ao} - P_e\right) = \frac{\rho_a U_{ae}^2}{2}$$

With:

$$0.92 < \eta_a < 0.96$$

$$U_{ae} = \frac{W_a}{\rho_a A(1-\tau)}$$

Let us note the conservation of momentum:

$$A\left(P_s - P_e\right) = W_a\left[-U_s\left(1+\sigma\right)\left(1+\lambda\right) + U_{ae} + \sigma U_{me}\right] \quad [1.2]$$

The coefficient λ denotes the loss of pressure in the mixing chamber.

On the other hand:

$$U_s = \frac{W_a\left(1+\tau\right)}{A\,\rho_r} \qquad \rho_r = \frac{W_a\left(1+\sigma\right)}{W_a\left(\frac{1}{\rho_a} + \frac{\sigma}{\rho_m}\right)} \quad [1.3]$$

After the mixing chamber, the diffuser obtains pressure recovery:

$$\left(P_r - P_s\right) = \eta_D \frac{\rho_r U_s^2}{2}\left(1-\delta\right) \; \delta < 1$$

η_D: yield of diffuser

$$0.75 < \eta_D < 0.8$$

δ: square of the cross-sections ratio of the divergent.

By replacing U_s with its value, we reach:

$$P_r - P_s = \frac{\eta_D}{2}\left(\frac{W_a}{A}\right)^2 \frac{\left(1+\sigma\right)^2}{\rho_r}\left(1-\delta\right) \quad [1.4]$$

Now, eliminate P_e between equations [1.1] and [1.2]:

$$P_{mo} - P_{ao} = \frac{1}{2}\left(\frac{W_a}{A}\right)^2 \left[\frac{\sigma^2}{\rho_m \tau^2 \eta_m} - \frac{1}{\rho_a (1-\tau)^2 \eta_a}\right] \qquad [1.5]$$

Eliminate P_s between [1.5] and [1.2]:

$$P_r - P_e = \left(\frac{W_a}{A}\right)^2 \left[\frac{(1+\sigma)^2}{\rho_r}\left(\frac{\eta_D (1-\delta)}{2} - (1-\lambda)\right) + \frac{1}{\rho_a (1-\tau)} + \frac{\sigma}{\rho_m \tau}\right] \qquad [1.6]$$

Eliminate P_e between [1.1] and [1.6]:

$$P_{mo} - P_r = \left(\frac{W_a}{A}\right)^2 \left[\frac{1}{2}\frac{\sigma^2}{\rho_m \tau^2 \eta_m} - \frac{\sigma}{\tau \rho_m} - \frac{1}{\rho_a (1-\tau)} - \frac{(1+\sigma)^2}{\rho_r}\left(\frac{\eta_D}{2}(1-\delta) - 1 - \lambda\right)\right] \qquad [1.7]$$

Assuming:

$$\Pi = \frac{P_{mo} - P_r}{P_{mo} - P_{ao}}$$

Eliminate $\frac{W_a}{A}$ by dividing expression [1.7] by [1.5]:

$$\Pi = \frac{\left[\frac{1}{2}\frac{\sigma^2}{\rho_m \tau^2 \eta_m} - \frac{\sigma}{\tau \rho_m} - \frac{1}{\rho_a (1-\tau)} - \frac{(1+\sigma)^2}{\rho_r}\left(\frac{\eta_D (1-\delta)}{2} - 1 - \lambda\right)\right]}{\frac{1}{2}\left[\frac{\sigma^2}{\tau^2 \rho_m \eta_m} - \frac{1}{\rho_a (1-\tau)^2 \eta_a}\right]} \qquad [1.8]$$

Note that $\Pi < 1$.

Knowing the three pressures, that is, Π, equation [1.8] allows for the calculation of σ according to Π and τ.

Equation [1.8] is the operational characteristic equation.

On the other hand, equation [1.5] or [1.7] provides the value of W_a/A, which determines the flows in terms of absolute value if we supply A.

We will perform the full calculation in the following hypotheses:

– Both fluids are of the same nature, in which case:

$$\rho_a = \rho_m = \rho_r$$

$$\eta_m = \eta_a = \eta_D = 1 \qquad \lambda = 0 \text{ and } \delta \# 0$$

The characteristic equation [1.8] is simplified:

$$\Pi = \frac{\left[\frac{1}{2}\frac{\sigma^2}{\tau^2} + \frac{(1+\sigma)^2}{2} - \frac{1}{(1-\tau)} - \frac{\sigma}{\tau}\right]}{\frac{1}{2}\left[\frac{\sigma^2}{\tau^2} - \frac{1}{(1-\tau)^2}\right]} \qquad [1.9]$$

If we need to solve this expression with respect to σ, we obtain a second degree equation:

$$\sigma^2\left[\frac{(\Pi-1)}{\tau^2} - 1\right] + 2\sigma\left[\frac{1}{\tau} - 1\right] + \left[-1 + \frac{2}{1-\tau} - \frac{\Pi}{(1-\tau)^2}\right] = 0$$

The discriminant of this equation is written as:

$$\Delta = \frac{(\Pi-1)^2 - \tau^4 + (1-\tau)^4}{\tau^2(1-\tau)^2}$$

If $\tau < 0.5$, which is typically the case, the discriminant Δ is positive, so we have:

$$\sigma = \frac{\frac{1-\tau}{\tau} \pm \sqrt{\Delta}}{\frac{1-\Pi}{\tau^2} + 1}$$

Only the symbol + corresponds with the real operation.

1.3. Gas ejectors and thermocompressors

1.3.1. *Parameters of the problem*

We use the one-dimensional theory which supposes that the properties of fluids and their behavior are constant through the area of their flow cross-section. A three-dimensional theory would account for these variations; however, the results of the one-dimensional theory are sufficient in accounting for experimental results.

We will take as known the results of compressible fluid mechanics in subsonic or supersonic flow.

Readers may refer to the work of Brun *et al.* [BRU 68].

The details are represented in Figure 1.2.

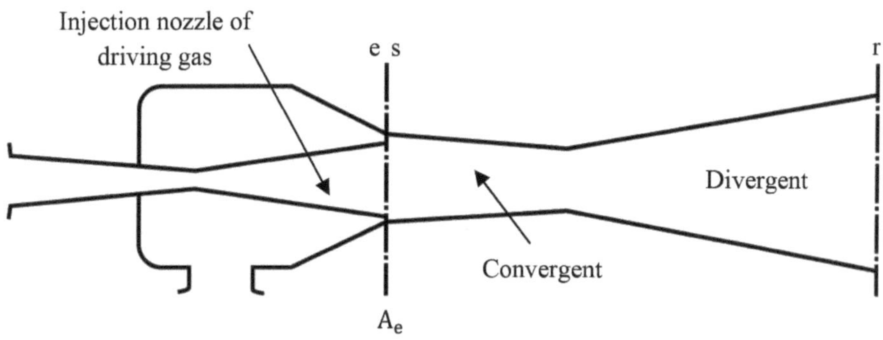

Figure 1.2. *Principle of a gas–gas ejector*

W_a: rate of suction gas ($kg.s^{-1}$)

A_e: cross-section of mixing passage (m^2)

G: flow quotient preceded by section A_e ($kg.s^{-1}.m^{-2}$)

τ_{col}: ratio of the nozzle neck cross-section to section A_e

τ_{tu}: ratio of the nozzle mouth cross-section to section A_e

σ: specific consumption (ratio of the driving fluid flow to the flow rate of suction fluid)

$V_{me,}$ V_{ae}: velocity of driving gas and suction gas prior to mixing ($m.s^{-1}$)

$\gamma_{m,}$ $\gamma_{a,}$ γ_r: ratio of specific heat for the driving gas, suction gas and the discharged mixture

M_m, M_a, M_r: molar mass of the driving gas, suction gas and the discharged mixture ($kg.kmol^{-1}$)

P_{ao}, T_{ao}: pressure generated and resting temperature of suction gas (Pa and K)

P_{mo}, T_{mo}: generated pressure and resting temperature of driving gas (Pa and K)

P_e: pressure prior to mixture (Pa)

P_s, T_s, ρ_s: pressure, temperature and density after mixture

P_r,T_r, ρ_r: pressure, temperature and density on discharge.

Densities are measured in $kg.m^{-3}$

The gas constant is $R = 8\ 314\ J.kmol^{-1}.K^{-1}$.

1.3.2. *Flow and velocity of driving fluid*

At the nozzle neck, the flow is sonic:

$$\frac{\sigma G}{\tau_{col}} = P_{mo}\left[\frac{\gamma_m M_m}{RT_{mo}}\left[\frac{2}{\gamma_m + 1}\right]^{\frac{1+\gamma_m}{\gamma_m - 1}}\right]^{0.5}$$

This expression is taken as the limit of the subsonic flow [BRU 68 p. 391] with support from the critical values [BRU 68 p. 389].

The ratio of the nozzle mouth cross-section to this neck is:

$$\frac{\tau_{tu}}{\tau_{col}} = \left[\frac{\left[\dfrac{2}{\gamma_m + 1}\right]^{2/(\gamma-1)} \dfrac{\gamma_m - 1}{\gamma_m + 1} \left[\dfrac{P_{mo}}{P_e}\right]^{2/\gamma_m}}{1 - \left[\dfrac{P_e}{P_{mo}}\right]^{(\gamma_m - 1)/\gamma_m}} \right]^{0.5} \qquad \text{[BRU 68 p. 408]}$$

Providing nozzle geometry, it is then possible to deduce the ratio P_e/P_{mo} for a supersonic flow by calculating successive approximations.

In the nozzle, relaxation is adiabatic:

$$\frac{T_{me}}{T_{mo}} = \left[\frac{P_e}{P_{mo}}\right]^{(\gamma_m - 1)/\gamma_m}$$

Also as with adiabatic relaxation, the velocity at the exit of the nozzle is obtained by:

$$V_{me} = \left[\frac{2\gamma_m R T_{mo}}{(\gamma_m - 1) M_m} \left[1 - \frac{T_{me}}{T_{mo}} \right] \right]^{0.5} \qquad \text{[BRU 68 p. 365]}$$

1.3.3. *Flow and velocity of the suction fluid*

The suction gas is sucked at pressure P_{ao} and brought to pressure P_e at the entrance of the mixing chamber.

The loosening ratio P_e/P_{ao} can be inferior or superior to the critical ratio r_{ac}.

$$r_{ac} = \left[\frac{2}{\gamma_a + 1}\right]^{\gamma_a/(\gamma_a - 1)} \qquad \text{[BRU 68 p. 389]}$$

1) $P_e/P_{ao} < r_{ac}$ supersonic loosening (see section 1.3.2):

Flow is calculated with the driving fluid:

$$\frac{G}{(1 - \tau_{tu})} = P_{ao} \left[\frac{\gamma_a M_a}{R\,T_{ao}} \left[\frac{2}{\gamma_a + 1}\right] \left[\frac{(1 + \gamma_a)}{\gamma_a - 1}\right] \right]^{0.5}$$

2) $P_e/P_{ao} > r_{ac}$ Subsonic loosening:

$$\frac{G}{(1-\tau_{tu})}) = P_{ao}\left[\frac{P_e}{P_{ao}}\right]^{\frac{1}{\gamma_a}}\left[\frac{2\gamma_a M_a}{(\gamma_a - 1)RT_{ao}}\left[1-\left[\frac{P_e}{P_{ao}}\right]^{\frac{\gamma_a - 1}{\gamma_a}}\right]\right]^{0.5} \quad \text{[BRU 68 p. 360]}$$

Temperature at the entrance of the mixing chamber corresponds to adiabatic loosening:

$$\left(\frac{P_e}{P_{ao}}\right)^{\frac{1}{\gamma_a}} = \frac{\rho_e}{\rho_{ao}} \qquad \frac{T_{ae}}{T_{ao}} = \left[\frac{P_e}{P_{ao}}\right]^{(\gamma_a - 1)/\gamma_a} \qquad \text{and} \qquad \frac{P}{\rho} = \frac{RT}{M}$$

The velocity V_{ae} of the suction fluid at the entrance of the mixing chamber is then obtained:

$$V_{ae} = \left[\frac{2\gamma_a RT_{ao}}{(\gamma_a - 1)M_a}\left(1 - \frac{T_{ae}}{T_{ao}}\right)\right]^{0.5} \quad \text{[BRU 68 p. 360]}$$

1.3.4. *Specific consumption σ*

This ratio is deduced from the outcome of the expressions that we have already calculated:

$$\frac{\sigma G}{\tau_{col}} \quad \frac{G}{1-\tau_{tu}}$$

Flow density G may also be deduced.

1.3.5. *Study of the mixing of two gases*

For the requirements of this calculation, we accept that the gases mix instantaneously as soon as the driving gas exits the nozzle. We assign indices as follows:

– index e for the gas before mixing (inlet);

– index s for the gas after mixing (exit mix);

– index r for the mix discharged on exit from the ejector.

The fundamental equations of the problem are as follows:

1) Conservation of mass and rule of mixtures.

As before:

$$W_m / W_a = \sigma$$

Concerning the molecular masses of the mixture, it should suffice to note that their kilo-mole number is conserved:

$$M_r = \frac{(1+\sigma)}{\dfrac{1}{M_a} + \dfrac{\sigma}{M_m}}$$

As for specific heat, this is the weighted average of the constituent heats:

$$C_{pr} = \frac{C_{pa} + \sigma C_{pm}}{1+\sigma} \quad \text{with} \quad C_p = \frac{\gamma R}{(\gamma - 1) M}$$

By replacing C_p with the value and resolving it in terms of γ_{r}, we reach:

$$\gamma_r = \frac{\dfrac{\gamma_a}{M_a(\gamma_a - 1)} + \dfrac{\sigma\gamma_m}{M_m(\gamma_m - 1)}}{\dfrac{1}{M_a(\gamma_a - 1)} + \dfrac{\sigma}{M_m(\gamma_m - 1)}}$$

The temperature T_r of the gas mixture at rest is obtained by writing out the sum of the enthalpies:

$$T_r = \frac{C_{pa} T_{ao} + \sigma C_{pm} T_{mo}}{(1+\sigma) C_{pr}}$$

2) Momentum equation:

$$A_e P_e + W_a (V_{ae} + \sigma V_{me}) = A_e P_s + W_a (1+\sigma) V_s$$

3) Energy equation:

We call the mass enthalpies I (J.kg^{-1})

$$\left(I_{ae}+\frac{V_{ae}^2}{2}\right)+\sigma\left(I_{me}+\frac{V_{me}^2}{2}\right)=(1+\sigma)\left(I_s+\frac{V_s^2}{2}\right)$$

Note that the enthalpy for gas at rest is written out as:

$$I_{mo}=I_{me}+\frac{V_{me}^2}{2}$$

$$I_{ao}=I_{ae}+\frac{V_{ae}^2}{2}$$

The enthalpies I_{mo} and I_{ao} are also perfectly defined.

On the other hand:

– the density after mixing is:

$$\rho_s=\frac{P_sM_s}{RT_s}$$

– the mass flow is then written as:

$$W_a(1+\sigma)=A_eV_s\times\rho_s=\frac{A_eV_sP_sM_s}{RT_s}$$

and the enthalpy at the end of mixing is written as:

$$I_s=C_{pr}T_s=\frac{\gamma_rRT_s}{(\gamma_r-1)M_r}$$

Ultimately, the energy equation can be written as:

$$W_a\left(I_{ao}+\sigma I_{mo}-(1+\sigma)\frac{V_s^2}{2}\right)=V_sA_eP_s\frac{\gamma_r}{\gamma_r-1}$$

We will now determine the flow properties of gas at the end of mixing.

The gas mixing is completely defined by its flow, which is known, and by its velocity V_s and its pressure P_s, which we will calculate.

1) Calculation of V_s, the velocity at the end of mixing:

The energy equation comprises two variables corresponding to the completion of mixing, i.e. V_s and P_s. P_s may be eliminated by means of the impulse equation:

$$A_e P_s = A_e P_e + W_a \left(V_{ae} + \sigma V_{me}\right) - W_a \left(1+\sigma\right) V_s$$

The second expression of the energy equation is written as:

$$\left[\frac{\gamma_r}{\gamma_r - 1}\right]\left[-V_s^2 \left(1+\sigma\right) W_a + V_s \left[A_e P_e + W_a \left(V_{ae} + \sigma V_{me}\right)\right]\right]$$

On the other hand, we can define an enthalpy I_r of mixing for the gases at rest:

$$I_r \left(1+\sigma\right) = I_{ae} + \sigma I_{mo} = \frac{\gamma_r R T_r \left(1+\sigma\right)}{\left(\gamma_r - 1\right) M_r}$$

The energy equation then becomes:

$$\frac{V_s^2}{2} W_a \left(1+\sigma\right)\left(1+\gamma_r\right) - V_s \gamma_s \left[A_e P_e + W_a \left(V_{ae} + \sigma V_{me}\right)\right] + W_a \left(1+\sigma\right) \gamma_r \frac{RT_r}{M_r} = 0$$

Multiply this expression by $\dfrac{2}{\left(1+\gamma_r\right) W_a \left(1+\sigma\right)}$

And we obtain:

$$V_s^2 - 2 V_s \gamma_r \frac{\left(A_e P_e / W_a + V_{ae} + \sigma V_{me}\right)}{\left(1+\gamma_r\right)\left(1+\sigma\right)} + \frac{2\gamma_r R T_r}{\left(1+\gamma_r\right) M_r} = 0$$

Assume that:

$$C_r^2 = \frac{2\gamma_r}{1+\gamma_r} \times \frac{RT_r}{M_r} \text{ (}C_r\text{: velocity of sound in the static mixture of the two gases)}$$

$$B = \frac{\gamma_r}{(1+\gamma_r)} \times \frac{(P_e / G + V_{ae} + \sigma V_{me})}{(1+\sigma)}$$

We can then calculate the mixing velocity exiting from the chamber by the following relation:

$$V_s^2 - 2\,B\,V_s + C_r^2 = 0$$

$$V_s = B - \left[B^2 - C_r^2\right]^{0.5}$$

2) Calculate the pressure P_s after mixing:

The impulse equation directly gives:

$$P_s = P_e + \frac{W_a}{A_e}\left[V_{ae} + \sigma V_{me} - (1+\sigma)V_s\right]$$

1.3.6. *Global compression ratio*

As soon as mixing is over, the gases enter the convergent–divergent. As the convergent is almost cylindrical, it can be accepted that, if V_s is subsonic, the flow will remain subsonic until it exits the device.

The speed of sound at the end of mixing is:

$$C_s = \left[\gamma_r \frac{P_s}{\rho_s}\right]^{0.5}$$

With:

$$P_s = G(1+\sigma) / V_s$$

γ_r and P_s are determined as before.

Typically, V_s is inferior to C_s.

Furthermore, we can accept that, on exiting the diffuser, velocity is zero. At this location, the pressure is the discharge pressure P_{ro}:

$$P_{ro} = P_s \left[1 + \frac{(\gamma_r - 1)\rho_s V_s^2}{\gamma_r 2 P_s} \right]^{\gamma_r/(\gamma_r - 1)} \qquad \text{[BRU 68 p. 369]}$$

Finally, the compression ratio is:

$$P_{ro}/P_{ao}$$

EXAMPLE 1.1.–

$P_{ao} = 0.25.10^5$ Pa $\quad P_{mo} = 9.10^5$ Pa $\quad \tau_{col} = 0.0694$

$T_{ao} = 339$ K $\quad T_{mo} = 448$ K $\quad \tau_{tu} = 0.364$

$M_a = M_m = 18 \quad \gamma_a = \gamma_m = 1.33$

$$\frac{\tau_{tu}}{\tau_{col}} = \frac{0.364}{0.0694} = 5.245$$

Hence (see section 1.3.2):

$$P_e = 0.20.10^5 \,\text{Pa}$$

$$\frac{T_{me}}{T_{mo}} = \left[\frac{0.20}{9} \right]^{0.248} = 0.389$$

$$V_{me} = \left[\frac{2 \times 1.33 \times 8314 \times 448}{0.33 \times 18} (1 - 0.389) \right]^{0.5} = 1010 \text{ m.s}^{-1}$$

$$\frac{T_{ae}}{T_{ao}} = \left[\frac{0.20}{0.25} \right]^{0.248} = 0.946$$

$$V_{ae} = \left[\frac{2 \times 1.33 \times 8314 \times 339}{0.33 \times 18} (1 - 0.946) \right]^{0.5} = 260.8 \text{ m.s}^{-1}$$

$$\frac{G}{1 - 0.364} = 0.25.10^5 \left[\frac{0.20}{0.25} \right]^{\frac{1}{1.33}} \left[\frac{2 \times 1.33 \times 18}{0.33 \times 8314 \times 339} \left[1 - \left[\frac{0.20}{0.25} \right]^{\frac{0.33}{1.33}} \right] \right]^{0.5}$$

Indeed, relaxation of the suction gas is subsonic:

$$\frac{0.2}{0.25} > \left[\frac{2}{2.33} \right]^{\frac{1.33}{0.33}} = 0.54$$

Hence:

$$G = 22.29 \text{ kg.s}^{-1}.\text{m}^{-2}$$

$$\frac{\sigma \times 22.29}{0.0694} = 9.10^5 \left[\left[\frac{2}{2.33} \right]^{\frac{2.33}{0.33}} \times \frac{1.33 \times 18}{8314 \times 448} \right]^{0.5}$$

$$(\text{see section } 1.3.2) \qquad \sigma = 4$$

$$B = \frac{1.33}{2.33 \times 5} \left[\frac{0.20.10^5}{22.29} + 260.8 + 4 \times 1010 \right] = 592.5 \text{ m.s}^{-1}$$

$$T_r = \frac{339 + 4 \times 448}{5} = 426 \text{ K}$$

$$C_r^2 = \frac{2 \times 1.33 \times 8314 \times 426}{2.33 \times 18} = 224632$$

$$V_s = 592.5 - \left[592.5^2 - 224632 \right]^{0.5} = 236.9 \text{ m.s}^{-1}$$

$$\rho_s = 22.29 \times 5 / 236.9 = 0.47 \text{ kg.m}^{-3}$$

$$P_s = 0.20.10^5 + 22.29(260.8 + 4 \times 1010 - 5 \times 236.9)$$

$$P_s = 89462 \text{ Pa}$$

$$P_{ro} = 89462\left[1 + \frac{0.33 \times 0.47 \times 236.9^2}{1.33 \times 2 \times 89462}\right]^{\frac{1.33}{0.33}}$$

$$P_{ro} = 103398 \text{ Pa} = 1.033 \text{ bar}$$

The compression ratio is then:

$$1.033 / 0.25 = 4.132$$

1.4. Practical applications of ejectors and thermocompressors

1.4.1. *Value of these devices*

Ejectors and thermocompressors contain no moving parts. Their immobility together with the simplicity of their constituent parts allows them to be made from a wide range of materials: metal, graphite, PVC and various resins, resulting in an appropriate resistance to corrosion.

In many cases, these devices can be built oversize without increasing the initial cost in order to allow for later increases in the capacity at little expense.

These devices can handle dust-containing gases.

Ejectors are typically used to maintain the vacuum in an installation.

Note that thermocompressors are of the same design as ejectors albeit considerably greater in size.

In practice, they are used to pressurize the water vapor obtained by vacuum evaporation for various purposes.

1.4.2. *Compression ratio and mounting ejectors*

This is the ratio P_{ro}/P_{ao}.

This ratio can reach 5 for an ejector and is less than or equal to 2 for a thermocompressor.

Typically, ejectors are placed in series and numbered according to the increasing pressure. Each ejector sucks in the discharge from the preceding one. The first ejector sucks within the installation and the last ejector discharges the highest pressure.

In order to give an approximation of the suction pressure that can be attained, a practical rule (assessed in mbar) is:

$$P_{ao} = 1013/6^n$$

where n is the number of ejectors in series.

If the driving gas is steam, it is advantageous to insert a condenser between two successive ejectors. In this manner, the next ejector does not need to suck the flow of steam, which has been condensed.

In general, the condensers used are of the mixing type and, occasionally, the surface type.

Figure 1.3 depicts two ejectors in series with an intermediate mixing condenser.

1.4.3. *Similarity between suction gases*

The form of the equations used shows that, for any given pressure, two parameters intervene:

$$\sqrt{\frac{T}{M}} \text{ and } \gamma$$

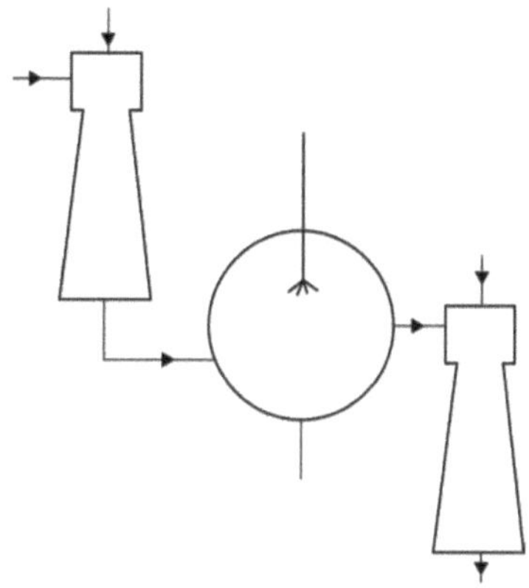

Figure 1.3. *Intermediate mixing condenser*

While these two parameters combine in a complex manner, they are not independent of one another. An approximate empirical correspondence can be established between the molecular mass ratio of the two gases and the suction rate of an ejector. Manufacturers provide these curves.

1.4.4. *Stability and stall point*

As can be observed through these calculations, a proportionality relationship exists between the pressures.

If the distribution network of the driving water vapor supplies high consumption appliances intermittent in operation, or indeed quite simply, if an accident occurs on the network, pressure P_{mo} can thus fall and, as a consequence, pressure P_{ro} can thus become inferior to the governing pressure of the device into which it is discharged. The direction of circulation in the convergent–divergent is inversed and the discharge empties by suction.

This phenomenon, known as counter-diffusion by manufacturers, occurs more readily if the specific consumption (that is, σ) is low. This can be explained by the fact that the water vapor jet does not have enough volume to occupy the entire mixing chamber.

1.4.5. *Ice formation at the exit of the ejection nozzle*

In the numerical example treated here, we note that by allowing the treated gas to be steam:

$$T_{me} = 0.397 \times 448 = 178\ K = -95°C$$

Consequently, ice formation can occur, particularly in small ejectors, and it is therefore important to heat the device when suction pressure is inferior to 5 Torr (1 Torr = 1 mm of mercury).

Of course, recompression occurs immediately after mixing, with temperature rising from T_{me} to T_s:

$$\frac{RT_s}{M_s} = \frac{P_s}{\rho_s}$$

1.4.6. *Regulation of ejectors*

Here, we cite three regulation methods listed by decreasing pertinence:

1) the best solution consists of injecting the water vapor to the first suction level, while the steam flow is regulated by a pneumatic valve, which is controlled by the suction pressure.

The advantage of water vapor condenses in the condenser after the first level without charging the subsequent levels.

Of course, instead of water vapor we could inject air or another gas. This gas could be of the same nature as the suction gas if it is taken at the exit of the final condenser;

2) we can also recycle a fraction of the discharge of the first level at the suction of this level, regulating it with a valve, as above;

3) finally, we can throttle the suction flow with a valve controlled by suction pressure. This method is only applicable if there is flow in the suction conduit. This method must be used with caution.

Generally, and particularly with high-volume installations, it is indispensable to obtain regulation with both a short response time and sufficient sensitivity to measure the pressure.

1.4.7. *Simplified calculation of specific consumption of an ejector*

1) Suction gases

When a distillation or evaporation installation works with a vacuum, air inevitably penetrates the device. The introduction of air occurs in the following two ways:

– air dissolved in the fluids brought to the installation;

– penetrating across gaskets, however impermeable they may be.

The air must then be extracted if the vacuum is to be maintained.

Often, the suction gases are a mixture of air and water vapor at the total absolute pressure P_a and temperature t_a. By knowing the air flow W_{aa}, it is easy to calculate the flow of water vapor W_{av} accompanying it.

$$W_{av} = W_{aa} \times \frac{18}{29} \times \frac{\pi(t_a)}{P_a - \pi(t_a)}$$

In this formula, we take the value 29 as the air molar mass.

The pressure π (t_a) is the saturation vapor pressure of water at the temperature t_a.

Generally, the equivalent suction flow W_{av} is given by:

W_{ai}: suction flow of constituent i (kg.h^{-1})

M_i: molar mass of constituent i (kg.kmol^{-1})

n: number of constituents, including air.

$$W_{av} = \sum_{i=1}^{n} W_{ai}\sqrt{\frac{18}{M_i}} \quad \left(\text{kg.h}^{-1}\right)$$

2) Specific consumption (empirical formula):

The specific consumption σ is the ratio of driving vapor flow rate the W_m to the suction rate W_a.

The calculation requires two parameters:

$$\rho = \frac{P_r}{P_a} \quad \text{and} \quad \mu = \frac{P_m}{P_a}$$

P_r and P_m are the absolute discharge pressure and the absolute pressure of the driving vapor. Parameter ρ is the compression ratio.

Specific consumption is given by:

$$\sigma = 0.74(\rho - 1)^{0.7} + \frac{11\rho^{1.786}}{\frac{230}{\rho^{1.89}} + \mu}$$

This formula is applicable for:

$$\mu > 20 \text{ and } 1.2 < \rho < 5$$

EXAMPLE 1.2.–

To resume the data of the theoretical example:

$$P_{mo} = 9 \text{ bar} \qquad P_{ao} = 0.25 \text{ bar} \qquad P_{ro} = 1.03 \text{ bar}$$

Hence:

$$\mu = \frac{9}{0.25} = 36 \qquad \rho = \frac{1.03}{0.25} = 4.12$$

$$\sigma = 0.74 \times 3.12^{0.7} + \frac{11 \times 4.12^{1.786}}{\frac{230}{4.12^{1.89}} + 36} = 4.3$$

This value is very close to that observed with the theoretical formulae as τ_{col} and τ_{tu} were given reasonable values.

1.4.8. *Conclusion*

The example provided here shows that the proposed empirical formula is applicable for water vapor sucking water vapor ($\gamma = 1.33$ and $M = 18$). This justifies, at least in part, the empirical correction formulae that we indicated for calculating W_a in the expression of P_{ro}, that means writing:

$$\rho_s V_s^2 = \frac{G_s^2}{\rho_s} = \text{cste} \ (\text{independent of the fluids' nature})$$

$$\text{Or even}: \frac{W}{\sqrt{M}} = \text{const.}$$

In reality, this simplified correction neglects the variations of γ (1.33 for water vapor and 1.4 for air) (see Appendix).

Recall that:

$$\gamma = \frac{C_p}{C_p - R}$$

when the molecular mass increases, C_p increases and γ decreases.

Finally, we should point out that the maximum compression rates generally accepted for ejectors are 5 and 2 for thermocompressors. Higher rates result in a prohibitive consumption of vapor.

2

Pipe Dimensions, Non-Newtonian Fluids, Liquid Hammer

2.1. Establishing pipe diameter

2.1.1. *Exterior diameter of metallic piping*

These diameters are identical worldwide for tooling investment reasons.

Non-professionals are often surprised to learn that these real diameters are quite different (particularly for small diameters) to the international nominal diameter (expressed in both millimeters and inches).

Table 2.1 gives some examples of pipes according to three parameters:

– real exterior diameter Φ_e in mm;

– the "ND" (nominal diameter in mm);

– the nominal diameter in inches (abbreviated: D"), with an inch equal to 25.4 mm.

NOTE.–

1) For velocity calculations, process engineers should systematically reduce the exterior diameters by 4 mm. Indeed, 2 mm is the minimum thickness of all tubes.

2) Unlike metallic tubes, plastic tubes are available with an interior diameter strictly equal to the ND, which helps the process engineer avoid errors.

Φ_e	10.2	13.7	17.2	21.3	26.8	33.40
DN	–	10 ?	–	15	20	25
D"	1/8	1/4	3/8	1/2	3/4	1
Φ_e	48.3	60.3	88.9	114.3	168.3	219.1
DN	40	50	80	100	150	200
D"	11/2	2	3	4	6	8
Φ_e	273	323.9	355.7	406.4	457	508
DN	250	300	350	400	450	500
D"	10	12	14	16	18	20
Φ_e	609.6	762	–	–	–	–
DN	600	750	–	–	–	–
D"	24	30	–	–	–	–

Table 2.1. *Pipe diameters*

2.1.2. *Choice of pipe velocity*

Several conditions and criteria apply when choosing pipe velocity:

– limits imposed by the process;

– implantation;

– particular properties of the treated fluid.

A. Process restrictions:

1) One essential criterion of a process is its functional stability. The installation pressure must vary slightly even if the flow value changes from simple to double. For normal flow, the solution consists of admitting moderated velocities that provoke only small drops in pressure. For gases, the typically accepted velocity remains in the order of 10 m.s^{-1}. For process liquids, levels of 1–1.5 m.s^{-1} are often accepted.

2) The frequency and duration of the operation periods also play a part. In an anti-fire installation, which by definition rarely operates and even then at most during half a day, very high water velocity levels of 3.5–5.5 m.s^{-1} may

be accepted. Thus, the purchase and installation cost of piping would be significantly reduced by an increase in the differential pressure of the pumps.

3) Clearly, wider diameters of pipes are required for greater flow. Whereas pressure drop calculations show that, at constant velocity and with constant friction coefficient, the pressure drop is inversely proportional to the diameter. Thus, in a turbulent flow regime, we can accept that velocity is proportional to the square root of the diameter. That is, for example:

1.5 m.s^{-1} in DN 100

3 m.s^{-1} in DN 400

This is why, in ventilation networks, we see velocities of 7–8 m.s^{-1} in the primary pipes (connected to devices) and 10–12 m.s^{-1} in main pipes, which are naturally greater in size.

B. Installation restrictions:

1) When a pipe is long, it entails a significant drop in pressure and requires a restriction of velocity, with another consequence being further increases in pipe costs (in order to avoid boiling a hot fluid, for example).

The importance of process restrictions typically decreases as the length of piping increases.

Indeed, we typically find:

– $L < 10$ m: the piping is entirely inside an elementary operation;

– 10 m $< L <$ 100 m: the piping is included in a manufacturing unit or installation;

– $L > 100$ m: for transportation between two different units of which at least one could be used for storage.

2) We must distinguish between naturally active lines and those equipped with a pressure-generating machine.

For example, a gravity flow would not be possible if the pressure drop is equal to the height distance, often 3–4 mLC (that is, meters liquid column).

On the contrary, process pumps are often capable of a discharge height of the order of 50 mLC.

In a turbulent flow regime, the ratio of possible velocities would be:

$$\frac{V_{gravity}}{V_{pump}} = \sqrt{\frac{3}{50}} = 0.25$$

Thus, we often accept 1.5–2 $m.s^{-1}$ on discharge from a pump and only 0.3–0.5 $m.s^{-1}$ in a gravity line.

3) Having defined the discharge line of a pump, it typically suffices to adopt the DN immediately superior for suction.

C. Particular properties of conveyed fluid:

1) If density is low, as in the case of low pressure steam, we can accept higher velocities reaching 40 and even 60 $m.s^{-1}$ for the following reason. Essentially, for a given flowrate, the pipe diameter increases if the density decreases. Thus given that, at a given velocity, the fall in pressure is inversely proportional to the diameter, the diameters for low pressure steam can reach values in the order of a meter.

2) If the viscosity exceeds 50 centipoises, the flow will certainly take the laminar regime and, the velocity must be reduced to 0.5 or 0.2 $m.s^{-1}$ (or even 0.05 $m.s^{-1}$ for a sludge) for pressure fall value to be moderated. Accordingly, some vigilance is required for iode solutions, mono- or polyethylene glycol solutions, food oils, petrol products, bitumen, etc.

3) However, high velocities can be accepted for water steam distribution networks (up to 50 $m.s^{-1}$) for two reasons:

– steam is under pressure and can consequently withstand a significant drop in pressure without provoking a significant relative variation in steam pressure;

– low-diameter piping exposes a reduced surface to the ambient temperature, which reduces thermal losses and condensation.

That means using the following expressions for the drop in line pressure:

a) wet steam:

$$\Delta P/(100 \text{ m}) = 0.015 \ P_v$$

where ΔP and P_v are measured in bars (P_v in relative bars)

b) if the steam is dry:

$$\Delta P/(100 \text{ m}) = 0.020 P_v$$

4) Utility water velocity in distribution networks typically ranges from 2 to 2.5 $m.s^{-1}$. Indeed, these relatively high velocities limit the tendency to coat the walls with either organic or inorganic material (microscopic algae). However, in continuous use, the velocity of 3 $m.s^{-1}$ cannot be exceeded without an increased risk of metal erosion.

5) If a salt or mineral suspension in a liquid is transported, in which the proportion of particles larger than 200 µm exceeds 30% of the mass, then velocity must be inferior to 2 $m.s^{-1}$ if unexpected decantation and blockages are to be avoided.

6) For certain fluids, friction with the pipe wall can generate static electricity, sparks and fire. This is why the velocity of carbon disulfide must always remain below 1 $m.s^{-1}$.

This is not the case for petroleum hydrocarbons, alcohols or even trichloroethylene, which may be conveyed at a velocity in the order of 2 $m.s^{-1}$.

2.1.3. *Available pressure drop*

Let us define the available pressure drop by:

$$\delta = \Delta P_F / (P_s - P_u)$$

ΔP_F is the pressure drop due to friction with the pipe wall. P_s and P_u (source pressure and working pressure) are the pressures at the ends of the pipe. P_s is often the discharge pressure from a pump or compressor.

In general:

$$0.1 < \delta < 0.2$$

Indeed, we must reserve the fraction $(1 - \delta)$ $(P_s - P_u)$ for line devices (among others: exchangers and particularly regulation valves).

2.1.4. *Pressure drop calculation*

In all systems, this is given by the formula:

$$\Delta P = f \frac{\rho V^2}{2} \frac{L}{D}$$

ρ: density of fluid ($kg.m^{-3}$)

V: mean velocity of fluid in the pipe ($m.s^{-1}$)

L: length of pipe (m)

ΔP: pressure drop (Pa).

Assume that:

$$f = \Psi$$

Ψ is then given by the Colebrook formula (see Brun *et al.* [BRU 68]):

$$\frac{1}{\sqrt{\Psi}} = 1.14 - 2\log_{10}\left[\frac{\varepsilon}{D} + \frac{9.32}{Re\sqrt{\Psi}}\right]$$

$$Re = \frac{VD\rho}{\mu}$$

μ: fluid viscosity (Pa.s)

ε: absolute roughness of the pipe (m) (see Table 2.2)

D: interior diameter of pipe (m)

Material	ε
Stainless steel and noble alloys	5.10^{-6}
Mild steel	50.10^{-6}
Concrete	500.10^{-6}

Table 2.2. *Examples of absolute roughness of pipes in different materials: these roughness values are simply orders of magnitude*

2.1.5. *Expression in practical units (fluids)*

$$\Delta P_F\left(\text{in bar}\right)=10^{-5}\,\Delta P_F\left(\text{in Pa}\right)$$

$$\rho_L\left(\text{in kg/m}^3\right)=10^3\,d_L\;\left(d_L:\text{density compared to water}\right)$$

$$D\left(\text{in m}\right)=10^{-3}\,D\left(\text{in mm}\right)$$

As a preliminary approximation we can estimate that:

$$f=0.03$$

On the other hand:

$$V\left(m.s^{-1}\right)=W\left(kg.h^{-1}\right)/\left[3600\times\frac{\Pi}{4}\times10^{-6}D^2\rho_L\right]$$

$$V\left(m.s^{-1}\right)=W\left(kg.h^{-1}\right)/\left[3600\times1000d_L\times\frac{\Pi}{4}\times10^{-6}\times D^2\left(mm\right)\right]$$

$$V=0.354\frac{W}{d_L D^2}$$

And, likewise:

$$\Delta P_F=\frac{1}{2}10^{-5}\times0.03\times10^3 d_L\times\left(0.354\right)^2\frac{W^2}{d_L^2 D^4}\times\frac{L}{10^{-3}D}$$

Hence:

$$D = 0.452\left[\frac{W^2L}{d_L \Delta P_F}\right]^{1/5}$$

EXAMPLE 2.1.–

$W = 14{,}000\ kg.h^{-1}$ $L = 50\ m$

$d_L = 1.2$ $\Delta P_F = 0.5\ bar$

$$D = 0.452\left[\frac{14000^2 \times 50}{1.2 \times 0.5}\right]^{1/5} = 50\ mm$$

$$V = \frac{0.354 \times 14000}{1.2 \times 50^2} = 1.65\ m.s^{-1}$$

2.1.6. *Gas*

Considering that the density of a gas compared with water is in the order of:

$$d_G = 10^{-3}\rho_G$$

the expression of D and V for fluids can be adapted for gases:

$$D = 1.8\left[\frac{W^2L}{\rho_G \Delta P_F}\right]^{1/5}$$

$$V = \frac{354\ W}{\rho_G D^2}$$

EXAMPLE 2.2.–

$W = 1{,}200\ kg.h^{-1}$ $L = 100\ m$

$\rho_G = 3\ kg.m^{-3}$ $\Delta P_F = 0.09\ bar$

$$D = 1.8\left[\frac{1200^2 \times 100}{3 \times 0.09}\right]^{1/5} = 100 \text{ mm}$$

$$V = \frac{354 \times 1.200}{3 \times 10^4} = 14.2 \text{ m.s}^{-1}$$

2.1.7. *Newtonian viscous fluids*

Bitumen is Newtonian above 65°C. A blowing operation, which is bitumen oxidation, increases its viscosity. Consequently, while we can make use of bitumen bases at 150°C directly from distillation due to its viscosity between 0.1–3 Pa.s, blown bitumen must be conveyed at above 180°C if we wish to enjoy similar levels of viscosity.

Bitumen viscosity according to the following law, behaves more or less so depending on temperature:

$$\mu/\mu_o = e^{-0.055(T - T_o)}$$

The viscosity of apple or grape juice as well as that of whole milk never exceeds 0.0025 Pa.s at 20°C. Similarly, most salt solutions, even when saturated, have twice the viscosity levels of water. These products can by no means be considered as viscous.

Conversely, the viscosities of olive oil and honey are respectively in the order of 0.085 and 5.5 Pa.s at 20°C.

The activated sludge from water treatment with 3% dry matter in mass have a viscosity at 20°C in the order of 0.1 Pa.s. However, if its dry matter content rises to 20%, viscosity is in the order of 1 Pa.s at 20°C.

Crystals dispersion in water at 10% of volume has an apparent viscosity below 0.005 Pa.s, as industrial crystals are often more than 100 µm in size. Such dispersions cannot be considered as viscous, and crystallization practitioners often make them similar to water.

Viscous products flow in laminar regime, for which the Reynolds number VDρ/μ is below 2100 and with the in pipe pressure drop given by:

$$\Delta P = \frac{32\mu VL}{D^2}$$

Also, inversely:

$$D = \sqrt{\frac{32\mu VL}{\Delta P}} = \left[\frac{128\mu WL}{\rho\pi\Delta P}\right]^{1/4}$$

ΔP: pressure drop (Pa)

μ: viscosity (Pa.s)

V: velocity ($m.s^{-1}$)

L: pipe length (m)

D: inner diameter of pipe (m)

P: density of product ($kg.m^{-3}$)

W: mass flow rate of product ($kg.s^{-1}$)

EXAMPLE 2.3.–

$W = 2.639\ kg.s^{-1}$ $\mu = 0.070\ Pa.s$ $\Delta P = 0.7\ bar = 0.7.10^5\ Pa$

$L = 50\ m$ $\rho = 900\ kg.m^{-3}$

$$D = \left[\frac{128 \times 0.07 \times 2.639 \times 50}{900 \times \pi \times 70.000}\right]^{1/4}$$

$$D = 0.05\ m$$

2.1.8. *Non-Newtonian fluids*

The behavior of these fluids is expressed by the following equation:

$$\tau = k\left[\frac{dV}{dr}\right]^n$$

V: local velocity of fluid ($m.s^{-1}$)

R: distance from pipe axis (m)

τ: shear stress (Pa)

k: consistence coefficient ($Pa.s^n$)

n: rheotypical exponent

If n = 1, we are dealing with a Newtonian fluid.

Sols (polyvinyl chloride suspensions in a plasticizer such as dioctylphthalate) are known as “dilatents” and in such cases, $n > 1$.

On the other hand, for food products, only products considered pseudo-plastic are below 1 ($h < 1$).

For example:

Product	Temperature (°C)	k ($Pa.s^n$)	n
Tomato concentrate 30%	30	19	0.4
Apricot purée 18%	27	5 – 20	0.3
Banana purée	20	110	0.33

Let us establish the relationship between flow and pressure drop. Shear stress on the wall is:

$$\tau = \frac{\pi r^2 \Delta P}{2\pi r L} = \frac{r}{2} \times \frac{\Delta P}{L} = k\left[\frac{dV}{dr}\right]^n$$

$$V(R) = \int_0^V dV = \left[\frac{\Delta P}{2kL}\right]^{1/n} \int_0^R r^{1/n} dr = \left[\frac{\Delta P}{2kL}\right]^{1/n} \left[\frac{n}{n+1}\right] R^{1+\frac{1}{n}}$$

The flow, in volume, is:

$$Q = \int_0^R V(R) 2\pi R dR = 2\pi \left[\frac{\Delta P}{2kL}\right]^{1/n} \left[\frac{n}{3n+1}\right] \left[\frac{n}{n+1}\right] R^{3+\frac{1}{n}}$$

Or, inversely:

$$\frac{\Delta P}{L} = 2k\left[\frac{Q}{2\pi}\right]^n \left[3+\frac{1}{n}\right]^n R^{-(1+3n)}$$

EXAMPLE 2.4.–

$V = 0.1 \text{ m.s}^{-1}$ $D = 0.156 \text{ m}$ $n = 0.4$

$L = 50 \text{ m}$ $k = 19$

$$Q = \frac{\pi}{4} \times 0.156^2 \times 0.1 = 0.00191 \text{m}^3/\text{s}$$

$$\frac{\Delta P}{L} = 2\times 19\left[\frac{0.00191}{2\pi}\right]^{0.4}\left[3+\frac{1}{0.4}\right]^{0.4}\left[1+\frac{1}{0.4}\right]^{0.4}\left[\frac{0.156}{2}\right]^{-(1+3\times 0.4)}$$

$\Delta P/L = 1.327 \text{ Pa.m}^{-1}$

$\Delta P = 1.327 \times 50 = 66{,}350 \text{ Pa}$

$\Delta P = 0.66 \text{ bar}$

2.2. Establishing pipe thickness

2.2.1. *Mechanical calculation conditions*

In order to calculate pipe thickness, we must first know not only the exterior pipe diameter Φ_e but also the two values known as mechanical calculation temperature and pressure, T_C and P_C.

The established practices in various countries mean that, for a given material and according to Φ_e, T_C and P_C, the minimum thickness required resists either internal overpressure or internal depression.

But a fluid circulating in a pipe is characterized by operating pressure and temperature. That is P_S and T_S for these two values. However, installations seldom operate in a rigorously stable manner and fluctuations in these values may occur (product flow rate, quality). In order to guard against these variations, operating conditions are increased. For example, we may accept as calculation values:

$$T_C = T_S + 15°C$$

And:

$$P_C = 1.1P_S + 1.4 \quad (P_S \text{ and } P_C \text{ in bar G})$$

Nonetheless, some processes can still be considered as processes at risk. This is the case of units including a heat complement or creation, in which the pressure of certain fluids increases due to saturation vapor. Such cases require a detailed safety study in order to bring to light extreme cases where pressure and/or temperature attains their maximum P_M and T_M values, clearly far above P_S and T_S. The temperature and pressure calculation would then be:

$$P_C = P_M \quad \text{and} \quad T_C = T_M$$

Of course, installation of release valves allows for the value of P_M to be reduced, which is not necessarily the case for that of T_M.

Finally, the calculation conditions will take the highest of the values deduced from these two processes.

NOTE.–

Mechanical calculations are made based on relative pressures (relative to the atmosphere), whereas process calculations and particularly the equipment specifications use absolute pressure. This frequently leads to unfortunate errors.

NOTE.–

Events accept a mechanical pressure calculation equal to their operating pressure, that is, 0 barG. There are even excesses. Temperature T_C is that of the device in question.

A gravity descent between two devices at atmospheric pressure is also calculated with, $P_C = 0$ barG. The calculation temperature is that of the device above.

A condensate descent into a pot is calculated with the P_C and T_C of the reception pot. This pot may have to be maintained in vacuum.

A gravity descent to the interior of a pressure process has a maximum pressure equal to that of the device above, augmented by the hydrostatic pressure.

2.2.2. *Pressure resistance*

The wall thickness is given by:

$$e = \frac{PD}{2(\sigma S + YP)}$$

Y: temperature-dependent coefficient (the value 0.4 is minimum)

P: overpressure relative to the exterior (MPa)

σ: permissible stress:

Steel at 20°C $\sigma = 138$ MPa

Stainless at 20°C $\sigma = 115$ MPa

S: welding coefficient:

Stretched unwelded tubes: 1

Rolled welded tubes: 0.8

The tubes can also be rolled without welding for diameters below a rather low limit.

D: exterior diameter of tube (m)

EXAMPLE 2.5.–

$$D = 114.3 \text{ mm} \qquad S = 1 \qquad \sigma = 138 \text{ MPa}$$

$$P = 55.2 \text{ bar} = 5.52 \text{ MPa} \quad Y = 0.4$$

$$e = \frac{5.52 \times 114.3}{2(138 + 0.4 \times 5.52)} = 2.28 \text{ mm}$$

2.2.3. *Vacuum resistance*

Cylinder thickness is given by:

$$e = 1.11 D \times \left[\frac{P}{E}\right]^{1/3}$$

P: negative pressure with respect to exterior (Pa, bar or MPa)

E: elasticity modulus (Pa, bar or MPa)

$$1 \text{ MPa} = 10^6 \text{ Pa} = 10 \text{ bar}$$

EXAMPLE 2.6.–

$$D = 0.6 \text{ m} \qquad E = 2.10^5 \text{ MPa} = 2.10^6 \text{ bar}$$

$$P = 0.06 \text{ bar}$$

Hence:

$$e = 1.11 \times 0.6 \sqrt[3]{\frac{0.06}{2.10^6}} = 0.002 \text{ m}$$

$$e = 2 \text{ mm}$$

NOTE.–

We recognize that the permissible stress taken for the pressure calculation (for fewer than 7,000 cycles between stress apparition and disappearance) is far below the elasticity modulus that is taken for the vacuum calculation. With the data of examples in sections 2.2.2 and 2.2.3:

$$\frac{\sigma}{E} \# \frac{115}{210^5} = 5.7510^{-4}$$

2.2.4. *Corrosion allowance and final thickness*

Subject to engineering practices, the corrosion allowance ranges from 1 to 1.5 mm for ordinary steel. It is absent for both noble alloys and stainless steel (and similarly for cast iron, cement and pipes in plastic materials). The deterioration of plastic materials is expressed by their swelling due to the infiltration of the process fluid.

Minimum thickness is obtained when the thickness required for mechanical resistance together with the corrosion allowance has been determined.

It should then suffice to search the manufacturer's catalogue for piping with the same exterior diameter and thickness immediately greater than or equal to the minimum thickness required (petroleum piping still employs the schedule concept, though this is purely conventional).

For example:

Φ_e	33.4	88.9
thickness	3.38	5.49

The schedule increases with the diameter Φ_e and the thickness increases with the schedule, thus with the diameter.

2.3. Flanges, seals and accessories

2.3.1. *Flanges and mounting*

The value of flange connections allows for easy dismounting for the maintenance or replacement of certain equipment.

We can distinguish the following modes of fixation:

– Weld neck flange: the neck of the flange is welded at the end of the tube. This is recommended for severe pressure and temperature conditions.

– Nested flange on the tube requires two weld beads. Easy to fit, though only suitable for moderate conditions.

– Compression flange slipped onto the tube. At its end, the tube has a neckler that is inferior in diameter to that of the flange. This neckler rather than the flange compresses the seal. This system is suitable for severe conditions.

– Screwed flanges (which only require a single weld) are used for DN below 50 mm though, as we will see, they are also used in systems with lenticular seals for very high pressure.

2.3.2. *The choice of flanges and seals*

A pair of flanges and the seal that separates them must remain leak-proof despite the pressure. In terms of leaks, flanges and seals represent pipes' weak points, since the pipe thickness is such that they never rupture under the influence of pressure (or vacuum).

The series of nominal pressures (N.P) stretches out from 2.5 to 1,000 bar and characterizes flanges, the thickness of which increases with the N.P.

The petroleum industry still refers to flanges of 180, 300, 600, 900, 1,500 and 2,500 pounds (that is: pounds per square inch (psi)).

The bearing side of flanges must be adapted to the type of joint selected:

1) Flat faces suit moderate temperatures and pressures. They are used on pipes in plastic materials associated with elastomer seals (Viton, Hypalon, ethylene-propylene, nitrile rubber) as tightening pressure is then limited. On metallic pipes for moderate conditions, they can be associated with armed elastomer seals and also with spiral seals.

2) The raised faces are used for moderately severe conditions. Armed elastomer joints are a requirement.

3) Simple-fitting and particularly double-fitting nested flanges are recommended for toxic fluids or fluids prone to leaking or for other fluids if the pressure is high enough or if a high vacuum is to be maintained. In this last instance, gas leakage by meter of seal perimeter remains below 0.5 $kg.h^{-1}$ measured in air. The faces of the nested flanges can be used with both reinforced elastomer seals and spiral seals.

4) Lenticular connection seals:

A plan for these connections is represented in Figure 2.1. This type of assembly is used for pressures in excess of 1,000 bars.

The lenticular seal is devised in a metal somewhat softer than that used in the pipe itself. Screw flanges are used here.

NOTE (Seals).–

Classic seals are made with armed polymers. Subject to the nature of the polymer, the seal should resist temperature variations, acidity and hydrocarbons.

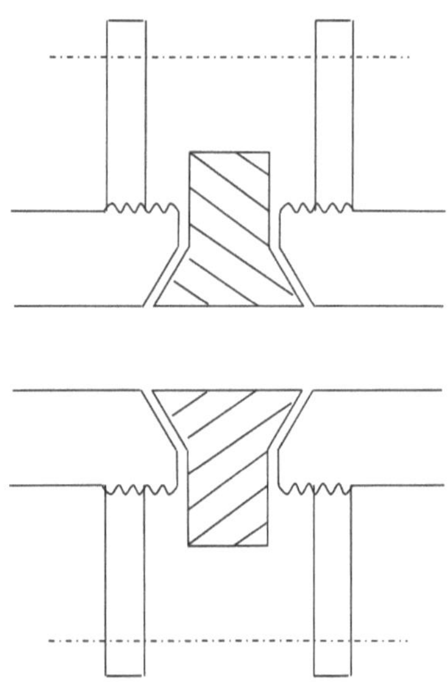

Figure 2.1. *Lenticular seal*

The calculation of admissible pressure is subject to pipe diameter and temperature:

DN \ T	40°C	150°C
100	130 bar	100 bar
400	100 bar	60 bar

If the seal does not perfectly withstand the conveyed fluid, the pressure limit above can be reduced. It would then be necessary to adopt spiral seals, which tolerate significantly greater pressures in normal use.

The durability of classic seals can be improved by coating them with a Teflon cover. This is not done with spiral seals due to the high variation in materials used, which means that they withstand all sorts of fluids.

For elastomer or armed elastomer seals, the absolute roughness ε of the flange faces must be such that:

$$25\ \mu m < \varepsilon < 50\,\mu m$$

On the other hand, for a spiral joint, we must have:

$$\varepsilon < 25\,\mu m$$

2.3.3. *Two types of pipe accessory*

A. Elbows:

Designers refer to the 3D model or 5D model as elbow due to the curvature radius of the tube being around 1.5D and 2.5D, respectively, where D is the exterior diameter of the pipe.

There are also elbows known as short radius, in which the curvature radius is equal to diameter.

B. Non-return valves:

We may distinguish:

1) Plastic ball valves:

A PVC ball on a PVC (polyvinyl chloride) seat with an EPDM.

These are employed for the conveyance of air or drinking water (hot and cold). They must be mounted vertically in upward direction.

2) Stainless steel ball valves:

The seat is PTFE. It can withstand 3 barG at 150°C and 10 barG at 50°C (which is its operating pressure).

3) Swing check valve:

They are used in horizontal pipes. A gate pivots on an axis. In closed position, it presses on its seat under the mere force of gravity, since the seat is slightly tilted vertically. Circulating in one direction, the fluid lifts the gate and can pass through it. In the other direction, the weight of the fluid presses the gate on its support, thereby blocking the passage.

Acceptable operating pressure is 16 barG at 100°C and 6 barG at 200°C. They are mounted on horizontal pipes. The obturator is made of stainless steel or bronze and the seal PTFE or armed elastomer. This valve is the most common and least expensive, but requires a stable flow of fluid.

If the pipe diameter is greater than or equal to 700 mm, double shutters are used, which reduce the inertia and intensity of the closing impact.

4) Spring valves:

The acceptable operating pressure is 40 barG at 120°C and 30 barG at 400°C. They are used for water, steam, condensates, air and vacuum networks. The fluid compresses a spring when it flows. They can be employed irrespective of the pipe orientation and are advised if the flow is unstable. The pressure drop varies according to the flow between 0.03 and 0.5 bar.

2.4. Sound waves in pipes

2.4.1. *Wave celerity*

When section A of the pipe is susceptible to vary slightly under the influence of the sound wave, the conservation equation is written as:

$$A\frac{\partial\rho}{\partial t}+\rho\frac{\partial A}{\partial t}+\rho A\frac{\partial u}{\partial x}+Au\frac{\partial\rho}{\partial x}+\rho u\frac{\partial A}{\partial x}=0$$

We can assume:

$$(\partial\rho/\partial P)_S=\rho\chi=1/C^{*2}$$

C^*: celerity of sound in the fluid supposed indefinite (not confined in a pipe) ($m.s^{-1}$)

(for water: $\chi = 4.92.10^{-10}$ Pa^{-1} at 20°C).

The continuity equation becomes:

$$\frac{1}{C^{*2}}\frac{\partial P}{\partial t}+\frac{\rho}{A}\frac{\partial A}{\partial t}+\rho\frac{\partial u}{\partial x}+\frac{u}{C^{*2}}\frac{\partial P}{\partial x}+\frac{\rho u}{A}\frac{\partial A}{\partial x}=0$$

Now, let us account for the elasticity modulus of the pipe material:

$$E=\frac{\sigma}{\Delta l/l}$$

(for steel, $E = 21.10^{10}$ Pa = 210 MPa)

Therefore, for a tube, the tensile-compressive stress σ is:

$$\sigma=\frac{DdP}{2e}$$

D: mean diameter of the pipe (m)

e: pipe thickness (m)

Hence:

$$\frac{dA}{A}=2\frac{dD}{D}=2\frac{dl}{l}=2\frac{\sigma}{E}=\frac{DdP}{eE}$$

That is:

$$\frac{1}{A}\left[\frac{\partial A}{\partial t}\right]_x=\frac{1}{A}\left[\frac{\partial A}{\partial P}\right]_x\left[\frac{\partial P}{\partial t}\right]_x=\left[\frac{\partial P}{\partial t}\right]_x\frac{D}{eE}$$

And:

$$\frac{\rho}{A}\frac{\partial A}{\partial t}=\frac{\rho\chi}{A\chi}\frac{\partial A}{\partial t}=\frac{1}{C^{*2}}\frac{\partial P}{\partial t}\frac{D}{\chi eE}$$

Equally:

$$\frac{\rho}{A}\frac{\partial A}{\partial x}=\frac{1}{C^{*2}}\frac{\partial P}{\partial x}\frac{D}{\chi eE}$$

Let us take these two expressions into the continuity equation:

$$\frac{1}{C^{*2}}\left[1+\frac{D}{\chi eE}\right]\frac{\partial P}{\partial t}+\frac{1}{C^{*2}}\left[1+\frac{D}{\chi eE}\right]u\frac{\partial P}{\partial x}+\rho\frac{\partial u}{\partial x}=0$$

This equation is identical to that of a plane wave in an undefined environment, provided it is accepted that:

$$\frac{1}{C^2}=\frac{1}{C^{*2}}\left[1+\frac{D}{\chi eE}\right]$$

Thus, the presence of the pipe wall leads to the multiplication of the celerity C^* of sound in the liquid supposed to be non-confined by a coefficient $\sqrt{k}$ such as the real celerity C in the pipe, that is:

$$C=\sqrt{k}C^*=\sqrt{\frac{k}{\chi\rho}} \quad \text{with:} \quad k=\left(1+\frac{D}{\chi eE}\right)^{-1}$$

EXAMPLE 2.7.–

Let us use the following physical data:

$$\chi = 4.9210^{-10}\ \text{Pa}^{-1} \quad \text{and} \quad \text{E} = 2110^{10}\,\text{Pa}$$

With, additionally:

$$\text{D}_\text{T} = 0.1\ \text{m} \quad \text{and} \quad \text{e} = 0.005\ \text{m} \qquad \rho = 1{,}000\ \text{kg.m}^{-3}$$

$$\text{C}_0 = \sqrt{\frac{1}{1000 \times 4.92.10^{-10}}} = 1.426\ \text{m.s}^{-1}$$

$$\text{D}_\text{T}/\text{e} = 0.100/0.005 = 20$$

The real celerity is:

$$\text{C} = 1426 \times \left[1 + \frac{20}{4.92 \times 21}\right]^{-1/2} = 1426 \times 0.915 = 1.305\ \text{m.s}^{-1}$$

2.4.2. *Natural damping of sound waves*

A priori, there are three natural causes possible for the damping of sound waves in fluids:

1) At a molecular level, the following effects combine, by increasing importance (see Temperley and Trevena [TEM 80]):

– thermal conduction;

– viscosity;

– molecular relaxation.

These effects are entirely negligible.

2) Friction on the pipe walls provokes damping detectable over distances in the order of a kilometer (see section 2.8.6).

3) The presence of microscopic bubbles of insoluble gas (air for example) is significant for damping detectable over several decimeters.

Then, we have:

$$\frac{P}{P_o} = e^{\frac{-Dt}{2}}$$

With:

$$D = \frac{C^2 F \alpha a M_G}{R \rho_G}$$

$$L = Ct$$

Readers may refer to a table of the values of α in the work of Temperley and Trevena [TEM 80, p. 113].

Thus, for benzene:

$$\alpha \# 8.4.10^{-12} f^2$$

where f is the frequency in Hz of the sound wave.

It requires:

$$\frac{Dt}{2} = 1.1 \quad \text{so that} \quad \frac{P}{P_o} = \frac{1}{3}$$

Hence:

$$L = \frac{2.2C}{D} = \frac{2.2 R \rho_G}{C F \alpha a M_G}$$

With:

$$F = \frac{\beta \rho}{x C_{pG}}; \qquad \text{so: } L = \frac{2.2 R \rho_G C_{pG} x}{C \beta \rho \alpha a M_G}$$

a is the surface of the bubbles for 1 kg of mixture:

$$a = \frac{6}{d_B} \frac{x}{\rho_G}$$

α is the thermal transfer coefficient which can be deduced from the Froessling formula:

$$Nu = \frac{\alpha d_B}{\lambda_L} = 2$$

d_B: bubble diameter (m)

λ_L: thermal conductivity of the fluid ($W.m^{-1}.K^{-1}$)

x: fraction of gas in mass

Therefore:

$$L = \frac{0.1833 R \rho_G^2 d_B^2 C_{pG}}{C\overline{\beta\rho}\lambda_L M_G}$$

The other intervening values are:

R: constant of ideal gas (8,314 $J.kmol^{-1}.K^{-1}$)

C_{pG}: mass thermal capacity of gas ($J.kg^{-1}.K^{-1}$)

$$\overline{\beta\rho} = x\beta_G\rho_G\left[\frac{\rho_E}{\rho_G}\right]^2 + (1-x)\beta_L\rho_L\left[\frac{\rho_E}{\rho_L}\right]^2$$

β_G and β_L: coefficients of thermal expansion volume of gas and fluid (K^{-1})

For an ideal gas: $\beta_G = \frac{1}{273} K^{-1}$ For water: $\beta_L = 0.2.10^{-3}\ K^{-1}$

ρ_E: density of emulsion ($kg.m^{-3}$)

$$\frac{1}{\rho_E} = \frac{x}{\rho_G} + \frac{1-x}{\rho_L}$$

x: mass fraction of gas

ρ_G and ρ_L: densities of gas and liquid ($kg.m^{-3}$)

M_G: molar mass of gas ($kg.kmol^{-1}$)

C: celerity of sound in emulsion:

$$\frac{1}{C^2} = x\left[\frac{\rho_E}{\rho_G C_G}\right]^2 + (1-x)\left[\frac{\rho_E}{\rho_L C_L}\right]^2$$

For a liquid: $C_L = \sqrt{\frac{1}{\rho_L \chi}}$ For a gas: $C_G = \sqrt{\frac{\gamma RT}{M_G}} = \sqrt{(\gamma - 1)C_{pG}T}$

γ: ratio of gas thermal capacities ($\gamma = C_P/C_V$)

EXAMPLE 2.8.–

Air in water:

$\rho_L = 1{,}000\ kg.m^{-3}$	$C_{pG} = 1{,}006\ J.kg^{-1}.K^{-1}$	$\chi = 4.92.10^{-10}\ Pa^{-1}$
$\rho_G = 1.29\ kg.m^{-3}$	$x = 0.000$	$\lambda_L = 0.5\ W.m^{-1}.K^{-1}$
$d_B = 10^{-2} m$	$M_G = 29\ kg.kmol^{-1}$	$T = 20°C = 293\ K$
$x = 0.0002$	$\gamma = 1.4$	$\beta_G = 3.663.10^{-3} K^{-1}$
		$\beta_L = 0.2.10^{-3} K^{-1}$

$$\frac{1}{\rho_E} = \frac{0.002}{1.29} + \frac{1}{1{,}000}$$

$$\rho_E = 865\ kg.m^{-3}$$

$$\overline{\rho\beta} = 0.0002 \times 1.29\left(\frac{865}{1.29}\right)^2 \times 3.66.10^{-3} + 0.2.10^{-3} \times 1000\left(\frac{865}{1000}\right)^2$$

$$\overline{\beta\rho} = 0.4245 + 0.1496 = 0.5741$$

$$C_G = \sqrt{\frac{1.4 \times 8\ 314 \times 293}{29}} = 343\ m.s^{-1} \qquad C_L = \sqrt{\frac{10^{10}}{1000 \times 4.9}} = 1{,}429\ m.s^{-1}$$

$$\frac{1}{C^2} = 0.0002\left[\frac{865}{1.29\times 343}\right]^2 + \left[\frac{865}{1000\times 1429}\right]^2$$

$$C = 36 \text{ m.s}^{-1}$$

$$L = \frac{0.1833\times 8314\times 1.29^2\times 10^{-4}\times 1006}{36\times 0.5741\times 0.5\times 29}$$

$$L = 0.85 \text{ m}$$

In conclusion, the presence of microscopic bubbles of insoluble gas very quickly absorbs the sound wave to the point that no liquid hammer effect is practicable. However, this situation is exceptional.

2.5. Mechanism of liquid hammer

2.5.1. *Closure of a valve, pressure equation*

The rapid closure of a valve is the prime cause in the apparition of the liquid hammer effect, which is expressed by a wave of over-pressure upstream from the valve.

When the fluid velocity in pipe section A_T differs from $\Delta u < 0$, the deficit in fluid mass (that has not passed through the valve), during time t, is:

$$-\rho A_T \Delta u t > 0 \qquad (\Delta u < 0) \qquad [2.1]$$

where ρ is the density of the fluid.

This mass of fluid has accumulated upstream, which provokes a compression, that is, a Δρ augmentation of density:

– for a column of fluid upstream of the valve, the volume of which is:

$$A_T(C - u_o)t$$

C: celerity of sound in the pipe ($m.s^{-1}$)

u_o: initial velocity of fluid in pipe ($m.s^{-1}$).

The volume of fluid arriving at the valve and just prior to passing through it:

$$A_T ut$$

In total, this represents the mass:

$$A_T t\Delta\rho(C+u-u_o)=A_T t\Delta\rho(C+\Delta u) \qquad [2.2]$$

By equalling the values [2.1] and [2.2], we obtain the conservation equation:

$$-\rho\Delta u = \Delta\rho(C+\Delta u) \qquad (\Delta u < 0) \qquad [2.3]$$

The disturbed fluid, having the same density $\rho + \Delta\rho$, occupies the total volume:

$$A_T(C+u-u_o)t = A_T(C+\Delta u)t$$

Therefore, its mass is:

$$(\rho+\Delta\rho)(C+\Delta u)tA_T = \rho CA_T t + A_T t\left[\Delta\rho(C+\Delta u)+\rho\Delta u\right]$$

The bracket is zero after equation [2.3] and the disturbed mass M is only:

$$M = \rho CA_T t$$

Hence, now ΔP is the over-pressure upstream of the valve. The impetus dissipated in this valve over time t was used to slow mass M of liquid, that is, to vary its velocity of $\Delta u < 0$:

$$A_T\Delta Pt = M(-\Delta u) = \rho CA_T t(-\Delta u) \qquad (\Delta P > 0 \text{ and } \Delta u < 0)$$

That is:

$$\Delta P = -\rho C\Delta u \qquad (\Delta u < 0) \qquad [2.4]$$

This equality is taken as the “pressure equation”.

Expression [2.4] represents the increase in pressure upstream of the valve. In the same fashion, we will show that a relaxation wave appears downstream of the valve with the same magnitude, equal to $+\rho C\Delta u$. We will generalize this.

Here, the x-axis will represent the distance along the pipe and, according to convention, the xs will increase from left to right. The initial velocity of the fluid, u_o, will be the algebraic value of this velocity counted on the x-axis.

The direction of wave propagation will be defined according to the direction of fluid propagation rather than according to the orientation of the x-axis, and we will define upstream and downstream according to fluid movement in the following way:

$\varepsilon_o = 1$ for a downstream wave (propagating in the same direction as the liquid).

$\varepsilon_o = -1$ for an upstream wave (propagating against the liquid).

In summary, we can write:

$$\Delta P = \varepsilon_o \rho C \Delta u$$

In this expression:

– celerity C is *always* positive;

– variation Δu is positive or negative.

For a compression: $\Delta P > 0$: for a relaxation: $\Delta P < 0$.

2.5.2. *Integrated pressure equation*

The pressure equation is:

$$\frac{dP}{du} = \varepsilon_o \rho C$$

This relationship is true locally and at a given time. However, accepting that ρ and C do not vary practically, this equation can be integrated:

$$p' = P - P_o = \varepsilon_o \rho C (u - u_o)$$

ρ: density of fluid ($kg.m^{-3}$)

C: celerity of sound in the pipe ($m.s^{-1}$)

u_0: velocity of liquid in the pipe prior to the event that has produced the liquid hammer ($m.s^{-1}$)

u: fluid velocity at the moment of sound wave ($m.s^{-1}$).

In most cases, the disturbance p' is provoked by valve closure.

EXAMPLE 2.9.–

$\rho = 1{,}000\ kg.m^{-3}$ $\quad C = 1{,}305\ m.s^{-1}$

$u_o = 2\ m.s^{-1}$ $\quad u = 0$

$$p' = P - P_o = 1000 \times 1305 \times 2 = 26.10^5\,Pa = 26\ bar\,.$$

Here, we may note the significance of over-pressure due to the liquid hammer effect.

2.5.3. *The complete equations of the problem*

We have described the liquid hammer on a basic level, which should be useful in illustrating the physical meaning of the phenomenon. Now, we will resume this treatment in a more mathematical light, explaining all of the equations of the problem.

At a given place (of abscissa x) and at a given time (or time t), velocity u of a liquid, its density ρ together with its pressure P must satisfy all of the following three equations. Here, velocity u is counted algebraically along the x-axis.

1) Equation of liquid state:

$$\frac{1}{\rho}\left[\frac{\partial \rho}{\partial P}\right]_S = \chi$$

With rapid changes in density, we may assume the absence of heat transport since heat transport is slow by nature. Thus:

– there is no heat transport from hot areas to colder areas;

– more generally, there is no decrease in compression energy due to internal irreversibility in the liquid.

It follows that derivative $(\partial \rho/\partial P)$ is taken at constant entropy S. Moreover, we note that χ is independent of temperature T for the limited domain of T variations of interest here.

The state equation can be integrated if, as we have just indicated, we accept that χ remains practically constant:

$$\mathrm{Ln}\frac{\rho}{\rho_o} = \chi\left(P - P_o\right)$$

2) The continuity equation

For a plane disturbance in an indefinite environment, this equation is written as:

$$\frac{\partial \rho}{\partial t} + u\frac{\partial \rho}{\partial x} + \rho\frac{\partial u}{\partial x} = 0$$

The state equation allows for the elimination of density in the conservation equation as:

$$\frac{\partial \rho}{\partial t} = \rho\chi\frac{\partial P}{\partial t} \quad \text{and} \quad \frac{\partial \rho}{\partial x} = \rho\chi\frac{\partial P}{\partial x}$$

Assuming:

$$\rho\chi = 1/C^2$$

As we know, this is the celerity of sound in fluid.

The conservation equation then becomes:

$$\frac{\partial P}{\partial t}+u\frac{\partial P}{\partial x}+\rho C^2\frac{\partial u}{\partial x}=0$$

3) The momentum equation is written as:

$$\frac{\partial u}{\partial t}+u\frac{\partial u}{\partial x}+\frac{1}{\rho}\frac{\partial P}{\partial x}=0$$

NOTE.–

By eliminating $\partial u/\partial x$ between the conservation equation and the impulse equation, we obtain:

$$\rho\frac{\partial u}{\partial t}+\frac{\partial P}{\partial x}=\frac{u^2}{C^2}\left[\frac{\partial P}{\partial x}+\frac{1}{u}\frac{\partial P}{\partial t}\right]=\frac{u^2}{C}\frac{dP}{dt}$$

We note in this equation that:

1) If a fluid is ideally incompressible, that is, if χ is non-zero and C is infinite, the preceding equation is reduced to:

$$\rho\frac{\partial u}{\partial t}+\frac{\partial P}{\partial x}=0$$

This is the classic impulse equation for in compressible fluids.

2) If the fluid is compressible and if we provoke a sudden variation in pressure (dP/dt high), the whole equation has to be employed, which is what we will work with.

3) If the disturbance in pressure is slow and if, moreover, we observe that u^2/C^2 is small $(2\ \text{m.s}^{-1}/1500\ \text{m.s}^{-1})^2$, we can use the impulse equation for incompressible fluids.

2.6. Approximate simulation method for simple pipes (without intermediate accessories and fittings)

2.6.1. *Assumption of mean values*

In this approximation, we accept that ρ and u do not deviate significantly from their initial value. In other words, ρ and u remain identical to ρ_o and u_o, respectively (this is clearly a subject of debate when velocity u gets from the value u_o to zero).

Accordingly, in this hypothesis, the momentum equation and the continuity equation are written as:

Momentum $$\tilde{n}_o \frac{\partial u}{\partial t} + \tilde{n}_o u_o \frac{\partial u}{\partial x} + \frac{\partial P}{\partial x} = 0$$

Continuity $$\frac{\partial P}{\partial t} + u_o \frac{\partial P}{\partial x} + \rho_o C^2 \frac{\partial u}{\partial x} = 0$$

Let us perform the following changes in variables:

$$\xi = u_o t + Ct - x \quad \text{and} \quad \eta = u_o t - Ct - x$$

In these expressions, the algebraic values of C and of u_o are supposed to be *of the same sign*. The derivation operations become:

$$\frac{\partial}{\partial x} = \frac{\partial \xi}{\partial x} \cdot \frac{\partial}{\partial \xi} + \frac{\partial \eta}{\partial x} \cdot \frac{\partial}{\partial \eta} = -\left[\frac{\partial}{\partial \xi} + \frac{\partial}{\partial \eta}\right]$$

Equally:

$$\frac{\partial}{\partial t} = (u_o + C)\frac{\partial}{\partial \xi} + (u_o - C)\frac{\partial}{\partial \eta}$$

The equations of the problem then become:

$$\left[\tilde{n}_o C \frac{\partial u}{\partial \hat{\imath}} - \frac{\partial P}{\partial \hat{\imath}}\right] - \left[\tilde{n}_o C \frac{\partial u}{\partial \text{ç}} + \frac{\partial P}{\partial \text{ç}}\right] = 0 \qquad \Big| -C \quad \Big| +C$$

$$C\left[\frac{\partial P}{\partial \xi}-\rho_o C\frac{\partial u}{\partial \xi}\right]-C\left[\frac{\partial P}{\partial \eta}+\rho_o C\frac{\partial u}{\partial \eta}\right]=0 \qquad \Bigg| +1 \quad \Bigg| +1$$

The linear combinations term by team give:

$$\frac{\partial}{\partial \xi}(P-\rho_o Cu)=0 \quad \text{and} \quad \frac{\partial}{\partial \eta}(P+\rho_o Cu)=0$$

Therefore, the grouping $(P - \rho_o Cu)$ is independent of ξ and depends only on $\eta = (u_o – C)\, t – x$.

Equally, the grouping $(P + \rho_o Cu)$ depends only on $\xi = (u_o + C)\, t – x$.

In other words, we can write where *F* is a function:

$$P=-\rho_o Cu+2F_{DS}\left[(u_o+C)t-x\right]$$

$$P=+\rho_o Cu+2F_{US}\left[(u_o-C)t-x\right]$$

Given that C and u_o have the same sign and that, in absolute values: $|C| >> |u_o|$, the grouping $(u_o – C)\, t – x$ characterizes a wave propagating in the opposite direction to that of u_o, that is, an upstream wave.

Equally, the grouping $(u_o + C)\, t – x$ characterizes a downstream wave.

Let us add and subtract these two equations member by member. We obtain (DS stands for downstream and US for upstream):

$$P=F_{AM}+F_{AV} \quad \text{and} \quad u=\frac{1}{\rho_o C}(F_{DS}-F_{US})$$

Thus, variables P and u are each the sum of the two disturbances:

–one upstream: $P=F_{US}$ $\qquad u=\frac{-F_{US}}{\tilde{n}_o C}$ [2.5]

–one downstream: $P=F_{DS}$ $\qquad u=\frac{F_{DS}}{\rho_o C}$ [2.6]

The initial conditions set the values F_{AM} and F_{AV} and, for t = 0, at all points for the given abscissa x, we have:

$$P_o = P(x,0) = F_{US} + F_{DS}$$

$$u_o = u(x,0) = \frac{1}{C\rho_o}(F_{DS} - F_{US})$$

Indeed, initial conditions are such that the entire length of Line P and u has a constant value.

2.6.2. *Physical interpretation*

By partial derivation of equations [2.5] and [2.6], we find the pressure equation:

$$\frac{\partial P}{\partial u} = \varepsilon_o \rho_o C$$

$\varepsilon_o = 1$ for a downstream wave

$\varepsilon_o = -1$ for an upstream wave.

Consequently, for a reduction in velocity ($du < 0$), it can only *climb* one compression wave ($dP > 0$) in the direction x decreasing and, only one relaxation wave ($dP < 0$) can *descend* toward the x increasing. This is the case irrespective of the limits conditions and irrespective of the u_o sign of initial liquid velocity.

To be more precise (with indices R for right and L for left):

– if $u_o > 0$, a relaxation is announced by < 0:

- F_R carries relaxation;
- F_L carries compression.

– if $u_o < 0$, relaxation is announced by > 0:

- F_R carries compression;
- F_L carries relaxation.

A valve set at the downstream end of the line, involving fluid slowing down, the above results in:

– all waves propagating downstream of the valve carry relaxation;

– all waves departing toward the upstream of the valve carry compression.

These conclusions would be inversed if u_o were negative.

All of the above are consistent with the basic description of the liquid hammer provided above.

2.6.3. *Boundary conditions*

1) Considerations on a capacity at abscissa x = 0:

Where P_R is the governing pressure of the capacity and P_0 the starting pressure of the pipe. The Bernoulli equation is:

$$P_R = P_0 + \frac{\rho u^2}{2}$$

Thus, with current data:

$$u = 2\ \text{m.s}^{-1} \qquad P_R = 0.1\ \text{bar} = 10{,}000\ \text{Pa} \qquad \rho = 1{,}000\ \text{kg.m}^{-3}$$

we find:

$$P_0 = 8{,}000\ \text{Pa}$$

The inclusion of the kinetic energy leads to a 20% difference in pressure in the steady regime P_R and P_0.

In the approximate method, that interests us here, and at abscissa x = 0:

$$P(0,t) = F_{DS} + F_{US}$$

$$P_R = F_{DS} + F_{US} + \frac{\rho u^2}{2}$$

$$u = \frac{1}{\rho_o C}\left(F_{DS} - F_{US}\right)$$

Eliminating u between the last two equations:

$$F_{AV}^2 + 2\left(\rho_o C^2 - F_{US}\right)F_{DS} + F_{AM}^2 + 2\rho_o C^2\left(F_{US} - P_R\right) = 0$$

Hence:

$$F_{AV} = F_{AM} - \rho_o C^2 + \rho_o C^2\sqrt{1 + \frac{2}{\rho_o C^2}\left(P_R - 2F_{US}\right)}$$

2) Closure of a valve at the right end:

Where P_∞ is the pressure downstream of the valve as, according to convention, the fluid always flows from right to left from the reservoir to the valve.

The valve characteristic is written as:

$$\Delta P = P(L,t) - P_\infty = \frac{1}{\eta^2(t)}\left[\frac{A_T}{A_{sm}^*}\right]^2 p\frac{\rho u^2}{2}$$

A_T: interior section of the pipe (m^2)

A_{sm}^*: minimum cross-section of the valve when completely open (m^2)

p: loss coefficient (non-recuperated) of kinetic energy after minimum cross-section

$\eta(t)$: valve partial opening coefficient

At time t, the minimum cross area is:

$$A_{sm} = A_{sm}^* . \eta(t)$$

Assume that, with a K_v other than in section 3.3.5:

$$K_v = \frac{1}{\eta^2(t)}\left[\frac{A_T}{A_{sm}}\right]^2 P$$

On the other hand:

$$P(L,t) = F_{AV} + F_{AM}$$

$$F_{DS} + F_{US} - P_\infty = K_V \frac{\rho_o u^2}{2}$$

$$u = \frac{1}{\rho_o C}(F_{DS} - F_{US})$$

Eliminating u in the previous equations:

$$F_{US}^2 - 2\left(F_{DS} + \frac{\rho C^2}{K_V}\right)F_{US} + F_{DS}^2 + \frac{2\rho_o C^2}{K_V}(P_\infty - F_{DS}) = 0$$

$$F_{US} = F_{DS} + \frac{\rho_o C^2}{K_V} - \frac{\rho_o C^2}{K_V}\sqrt{1 - \frac{K_V}{\rho_o C^2}(P_\infty + 2F_{US})}$$

2.6.4. *Propagation diagram*

The propagation lines are represented using two coordinate axes:

– horizontal, distance x along the line;

– vertical, time t.

These curves are called propagation *characteristics*.

Propagation to the right corresponds to:

$$t - \frac{x}{C_{DS}} = K_{DS} = \text{const.}$$

And to the left:

$$t + \frac{x}{C_{US}} = K_{US} = \text{const.}$$

From the above:

To the right $F_{AV}\ (K_{DS}) = \text{const.}$

To the left $F_{AM}\ (K_{US}) = \text{const.}$

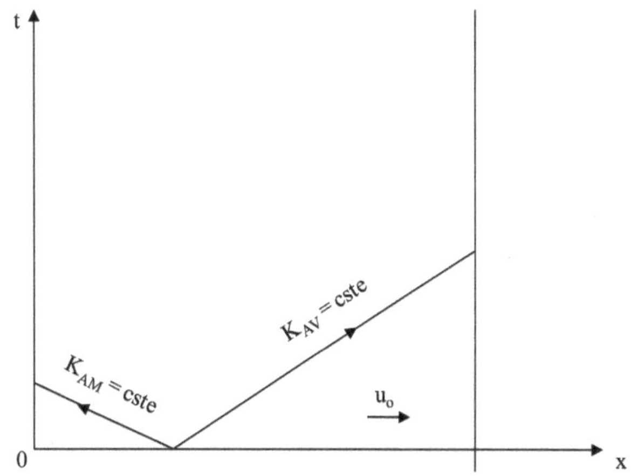

Figure 2.2. *Propagation diagram (principle)*

For t = 0, values u and P along the x-axis define the initial condition.

For x = 0 or x = L, values u and P depending on time define the boundary conditions.

We have divided the characteristic function into areas that will be described below (Figure 2.3):

– Area 0: a combination (superposition) of upstream levotropic wave levotropes (that propagate to the left) and downstream dextrotropic waves (that move toward the right). These waves arise from the initial condition along the line.

– Area 1: the dextrotropic waves move from the point of origin according to the passage of time. They combine with levotropic waves coming along the line.

– Area 2: in a symmetrical way with area 1, the levotropic waves move from the right end combining with the dextrotropic waves coming along the line.

– Area 3: the dextrotropic waves from the origin combine with the levotropic waves coming from the right end.

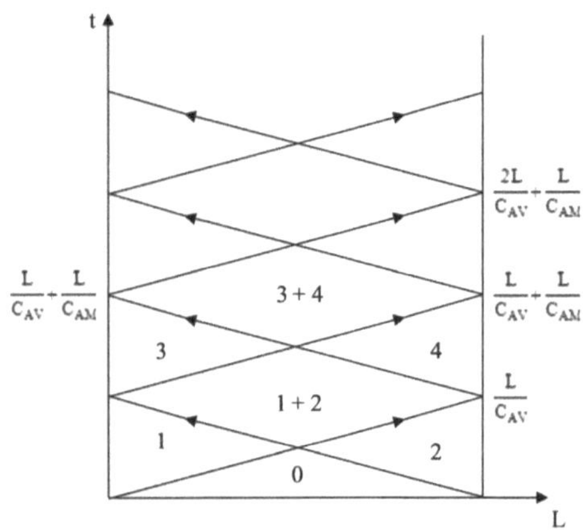

Figure 2.3. *Characteristic function of propagation*

Finally, the central lozenges combining the two waves F_{US} and F_{DS} whose number of reflections and the ends of the lines are equal, the right triangles (2, 4, 6, etc.) and the left triangles (1, 3, 5, etc.) combine two waves that have undergone a number of reflections whose difference is 1.

2.6.5. *Calculation procedure*

We use a reduced unit of time equal to $\frac{tC}{2L}$, where L is the length of the line.

Time 2L/C corresponds to a sound wave round-trip.

We then assess the closure time τ_f of the valve in reduced units.

The initial conditions give F_{AVi} to the right of the valve.

For the first increment:

$$K_V = \frac{1}{\eta^2(\Delta\tau)}\left[\frac{A_T}{A_{sm}}\right]^2 P$$

And:

$$F_{US} = F_{DS} + \frac{\rho_o C^2}{K_V} - \frac{\rho_o C^2}{K_V}\sqrt{1 - \frac{K_V}{\rho_o C^2} P_\infty}$$

The pressure wave F_{US} climbs along the line and is reflected consistent to the reflection equation on the process device. The characteristic function resembles Figure 2.3 and announces successive reflections at the ends of the.

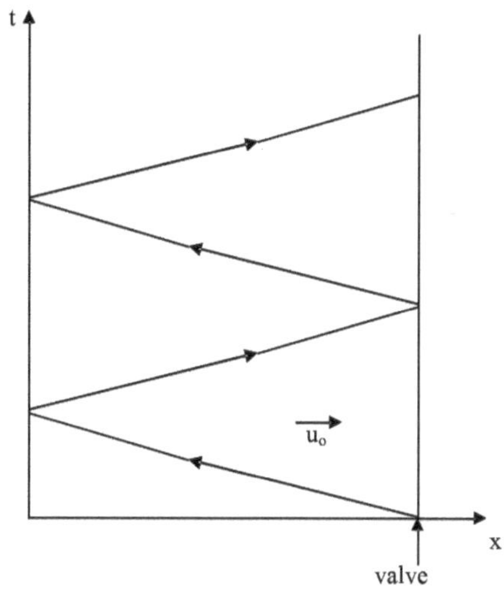

Figure 2.4. *Compression wave (valve closure)*

In his book, Moody [MOO 90, pp. 378–379] gives the results for the calculation of successive reflections for the areas defined in Figure 7.11 of his work as a function of the time variations of the boundary conditions.

The same author chose the example of a valve that closes according to the rule:

$$\eta(t)=\frac{A_{sm}}{A^{*}_{sm}}=1-\left[\frac{t}{\tau_f}\right]^n$$

τ_f is the closure time and section A^{*}_{sm} corresponds to the place in the valve where cross-section is minimum (that is, in the place of its obturator, when the valve is wide open).

The calculation results obtained by this author are as follows (see his Figure 7.16):

– n = ½: the closure is *slowed* toward the end – P (L,t) passes a maximum equal to around ¼ of F_{DS}.

– n = 1: closure is proportional to time – P (L,t) follows a slow development in flow that then accelerates to finally stabilize at a threshold with a level in the order of $F_{AV}/2$.

– n = 2: closure is *accelerated* toward the end – P (L,t) has a value equal to approximately 1.3 F_{AV}.

We note that, if a reduction in the pressure effects is required, the valve must be closed more and more slowly as total closure is reached.

2.7. Simplified graphic method

2.7.1. *Preliminary*

We will apply this method to the typical line comprising a tank to the left and a block valve to the right that is to be closed.

We will show why it is advantageous to close the valve slowly to avoid unnecessary wear on the equipment.

The method is based on several hypotheses:

– u is disregarded in favor of C;

– C is constant;

– in order to describe the reflections at the line ends, we do not apply the Bernoulli equation, that is, the kinetic energy of the fluid is disregarded.

At the tank, we simply write the characteristic equation of the device:

$$P = P_R = \text{const.}$$

Reflection in the reservoir retains velocity and transforms the pressure disturbance into its opposite.

At the valve, the characteristic equation is (see section 2.6.3):

$$P - P_\infty = K_V \frac{\rho u^2}{2} \quad \text{with} \quad K_V = \frac{p}{\eta^2(t)} \left[\frac{A_T}{A_{sm}} \right]^2$$

where P_∞ is the pressure downstream of the valve.

2.7.2. *Practical procedure*

It is convenient to use a reduced unit of time equal to tC/2L for the graph, where L is the line length. Time 2L/C corresponds to one return trip of the sound wave.

In addition, it is assumed that the valve closure occurs in discontinuous, instantaneous and successive increments.

The pressure equation is:

$$\frac{dP}{du} = +\rho C \text{ for a wave moving downstream}$$

$$\frac{dP}{du} = -\rho C \text{ for a wave moving upstream}$$

Let us take the following coordinate system:

– pressure P on the vertical axis;

– velocity $u = Q/A_T$ on the horizontal axis.

Assume that the characteristic equation of a valve is written as follows:

$$\Delta P = \frac{p}{\eta^2(t)}\left[\frac{A_T}{A_{sm}}\right]^2 \frac{\rho u^2}{2}$$

The result is a parabola that progressively approaches the vertical as the valve closes, that is to say, $\eta(t)$ moves toward zero.

Let us suppose that the closure occurs in five units of reduced time. Characteristic curve (5), corresponding to $\eta(t) = 0$, merges with the vertical axis.

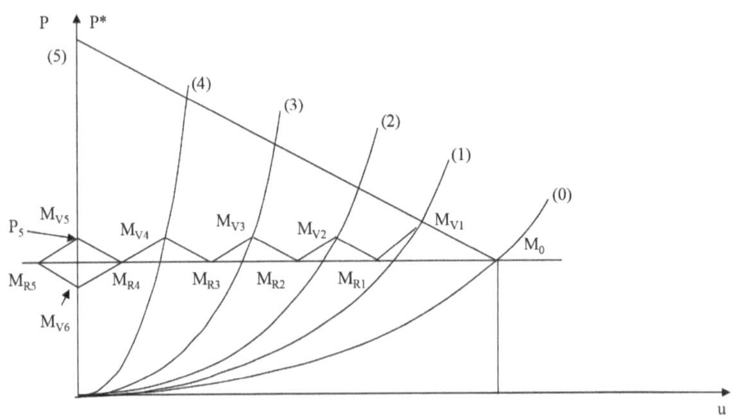

Figure 2.5. *Development of pressure on valve closure*

The initial condition corresponds to point M_0 (P_0, u_0).

Now, we will consider the process cycle:

a) Flow 0 (initial):

u_0 and P_0

b) Wave 0 (from the valve to the tank):

The valve characteristic goes from configuration (0) to configuration (1).

$$-\Delta u_1 \quad \text{and} \quad \Delta P_1 = -\rho C(-\Delta u_1)$$

c) Flow 0.5 (total of flow 0 and wave 0):

$$u_0 - \Delta u_1 \quad \text{and} \quad P_0 + \Delta P_1$$

These are the coordinates from the M_{V1} intersection point of characteristic (1) with the right passing by M_0 and gradient $-\rho C$.

d) Wave 0.5 (from the reservoir toward the valve):

This is deduced from wave 0 by adopting the same velocity value and an opposite value for pressure:

$$-\Delta u_1 \quad \text{and} \quad -\Delta P_1$$

e) Flow 1:

This is the total of flow 0.5 and wave 0.5:

$$u_0 - 2\Delta u_1 \quad \text{and} \quad P_0 \quad (\text{coordinates of point } M_{R1})$$

Point M_{R1} is the intersection of the vertical $P = P_0$ with the right line passing by M_{V1} and slope $+\rho C$.

f) Wave 1 (from the valve toward the tank):

From flow 1, the valve characteristics change to configuration (2) with the release of compression wave 1.

$$-\Delta u_2 \quad \text{and} \quad \Delta P_2 = -\rho C(-\Delta u_2)$$

g) Flow 1.5:

This is the total of flow 1 and wave 1.

$$u_0 - 2\Delta u_1 - \Delta u_2 \quad P_0 + \Delta P_2 \quad (\text{coordinates of point } M_{V2})$$

Point M_{V2} is the intersection of characteristic (2) with the right line passing by M_{R1} and slope $-\rho C$.

M_{V5} depicts condition 5 of the system $(0, P_0 + \Delta P_5)$ for complete valve closure. It is condition 5 that corresponds to the propagation of a 4.5 levotropic wave $(-\Delta u_5, \Delta P_5)$.

This wave is reflected on the reservoir to give dextrotropic wave 5 $(-\Delta u_5, -\Delta P_5)$ and condition 5.5 is the total of wave 5 and condition 5:

$$-\Delta u_5, \, P_0 \quad (\text{point } M_{R5})$$

From condition 5.5, the valve characteristics (equated with the vertical axis) give point M_{V6} for a levotropic wave and, ultimately, the point representation system indefinitely follows the peaks of the diamonds:

$$M_{R4}, \, M_{V5}, \, M_{R5}, \, M_{V6}$$

Consequently, the phenomenon becomes intermittent with a duration equal to that of two return-trips, finally weakening due to the mechanisms of piping friction (see section 2.8.6).

In conclusion, the successive round-trips of the wave between the valve and the reservoir are depicted by the broken line M_0, M_{V1}, M_{R1}, M_{V2}, M_{R2}, etc.

Pressure P_5 to the right of the valve, emitted toward the left, differs little from P_0 and the five successive reflections lead to a moderate liquid hammer wave amplitude.

On the other hand, where the valve is closed abruptly, the intervening characteristic is that of the vertical axis. Pressure would be P^*, which is far greater than P_5.

In the first instance, the closure time was equal to 10 L/C and, in the second instance, was far shorter, inferior to L/C.

This graphical method, while approximate, is a simple illustration of a known phenomenon, showing that a valve must be closed slowly in proportion to the length of the pipe it controls. Typically, the closure time in the order of 5 s avoids negative liquid hammer consequences.

2.8. Anti-liquid hammer chambers

2.8.1. *Definition*

These devices minimize over-pressurization without themselves being subject to excessive fluid-level oscillations.

As the duration of sound waves is very short in comparison to that of the fluid movement, the separation of these effects is possible.

2.8.2. *Influence of the chamber on over-pressure*

We will consider two cases: where the tank is located immediately before the valve and where it is much further upstream.

1) If the tank is immediately upstream of the valve:

The pressure disturbance in the pipe due to total and immediate valve closure would be:

– in the absence of a tank:

$$P'^{*} = \rho C u_o$$

– with the presence of a chamber:

$$P' = \rho C(u_o - u_1) \qquad [2.7]$$

In fact after closure, fluid continues to flow through the pipe to fill the chamber at velocity u_1.

However, this acute pressure disturbance provokes an increase in tank level (at cross-section A_B) at velocity V_{B1}, so that:

$$V_{B1} A_B = u_1 A_T$$

Velocity inside the tank immediately goes from zero to the value V_{B1}, which corresponds to the preceding pressure disturbance p:

$$P' = \rho C V_{B1} = \rho C \frac{A_T}{A_B} u_1 \qquad [2.8]$$

By eliminating p' between [2.7] and [2.8], we have:

$$\frac{u_1}{u_o} = \frac{1}{1 + \frac{A_T}{A_B}}$$

Indeed, the "damping" (below 1) due to the presence of the tank is expressed by:

$$\frac{P'}{P'^*} = \frac{A_T}{A_B}\frac{u_1}{u_o} = \frac{1}{1 + \frac{A_B}{A_T}}$$

2) If the tank is located somewhere upstream, the pressure transmitted upstream of the tank can be written as:

$$\frac{P_t^{'}}{P'^*} = \frac{P_t^{'}}{P_i^{'}} = \frac{2A_T}{2A_T + A_B} = \frac{1}{1 + \frac{A_B}{2A_T}}$$

The reduction is less pronounced.

3) It is possible to show that a restriction at the entrance of the liquid hammer tank results in diminished loss.

2.8.3. *Fluid-level oscillations inside the tank*

Let us assume that the chamber is located immediately before the valve. The connecting pipe is free of any restriction that could lead to a reduction in liquid movement.

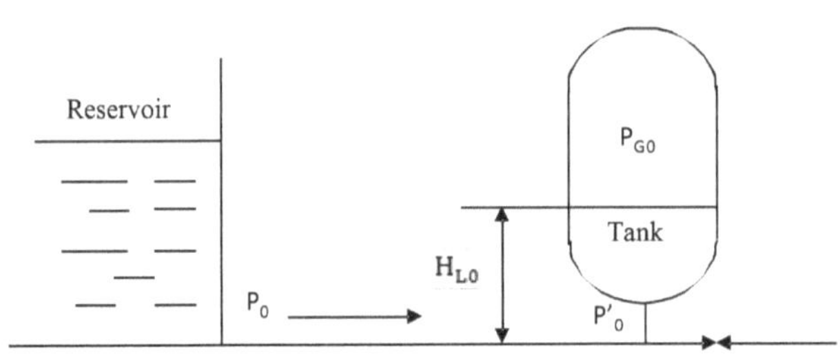

Figure 2.6. *Placement of liquid hammer tank*

In steady regime:

$$P_0 - P_0' = f\frac{\rho V_0^2}{2}\frac{L}{D} = P_0^*$$

We assume that the tank "gas sky" is in permanent at pressure P_{G0} and occupies height $H_{GO} = H_B - H_{L0}$.

We must have:

$$P_0' = P_{G0} + \rho g H_{L0}$$

On valve closure, the pressure immediately upstream of the valve takes the following value according to the rise z of fluid in the tank:

$$P_1' = P_G + \rho g\left(H_{L0} + z\right)$$

The gas is reduced in volume and pressure by isothermal compression (rather than isentropic, as the gas is in contact with the fluid, whose thermal capacity is far greater):

$$P_G = P_{G0}\frac{\left(H_B - H_{L0}\right)}{\left(H - H_{L0} - z\right)}$$

So that:

$$P_1' = P_{G0}\frac{\left(H_B - H_{L0}\right)}{H_B - H_{L0} - z} + \rho g\left(H_{L0} + z\right)$$

Where Q is the flow of fluid entering the tank, and A_B and A_T are the respective cross-sections of the chamber and the pipe, the conservation of fluid is written as:

$$Q = A_B\frac{dz}{dt} = A_T\left(V_T - V_V\right) \qquad [2.9]$$

V_T: velocity of fluid in the pipe ($m.s^{-1}$)

V_V: velocity of fluid between the tank and the valve in case the latter is not completely closed (m.s^{-1}).

For the cylinder of fluid present in the pipe between the reservoir and the tank (whose volume is $A_T L$), the dynamic equation is written as:

$$L\rho\frac{dV_T}{dt} = (P_0 - P_1) - (P_0 - P_{10}) + \varepsilon P^*$$

That is:

$$\rho L\frac{dV_T}{dt} = P_{G0}\left[1 - \frac{H_B - H_{L0}}{H_B - H_{LO} - z}\right] - \rho g z + \varepsilon P^* \quad [2.10]$$

The force arising due to friction and expressed using P^* is always opposite in direction to V_T. Hence:

$\varepsilon = -1$ if V_T is directed toward the chamber ($V_T > 0$)

$\varepsilon = 1$ if V_T is directed toward the tank ($V_T < 0$)

In finite differences, the equations [2.9] and [2.10] are written as:

$$\Delta z = \frac{A_T}{A_B}(V_T - V_V)\Delta t$$

$$\Delta V = \left[\frac{-gz}{L} + \varepsilon\frac{\rho V^2}{2D} - \frac{P_{G0}z}{H_B - H_{L0} - z}\right]\Delta t$$

The independent variable is time t for which the increments are given to be 1–3 s.

Ultimately, this concerns periodical damped oscillations characterized by:

– potential energy (fluid level and gas pressure);

– kinetic energy (fluid velocity in pipe);

– nonlinear damping due to friction in the pipe.

We note that only in the presence of an anti-liquid hammer tank is pipe friction effective to reduce the liquid hammer wave.

2.8.4. *Pipe friction equations*

The equation of sound without friction can be obtained from the conservation equations so long as the terms $\partial P/\partial x$ and $u\partial u/\partial x$ are neglected and momentum. The equations are then written as:

$$\frac{\partial P}{\partial t}+\rho C^2\frac{\partial u}{\partial x}=0 \qquad [2.11]$$

$$\frac{\partial u}{\partial t}+\frac{1}{\rho}\frac{\partial P}{\partial x}=0 \qquad [2.12]$$

If we derive [2.11] in respect of t and [2.12] in respect of x, we obtain:

$$\frac{\partial^2 P}{\partial t^2}+\rho C^2\frac{\partial^2 u}{\partial x\partial t}=0 \qquad [2.13]$$

$$\frac{\partial^2 u}{\partial x\partial t}+\frac{1}{\rho}\frac{\partial^2 P}{\partial x^2}=0 \qquad [2.14]$$

We shall add equations [2.13] and [2.14] member by member after multiplication of [2.13] by $-\rho C^2$

$$\frac{\partial^2 P}{\partial t^2}-C^2\frac{\partial^2 P}{\partial x^2}=0$$

We can easily check that the solutions to this equation are of the form:

$$P=P\left(t\pm\frac{x}{C}\right)$$

These are waves that propagate at velocity C toward the left or toward the right.

We note that the vibrating string approximation leads to a small error as to the propagation velocity: which is C rather than $C \pm u_o$.

With friction, the equations of the problem are written as:

– continuity: $$\frac{\partial P}{\partial t} + ñC^2 \frac{\partial u}{\partial x} = 0$$

– momentum: $$\frac{\partial u}{\partial t} + \frac{1}{\rho}\frac{\partial P}{\partial x} + Ru = 0$$

with: $$R = \frac{f\overline{|u|}}{2D}$$

f: friction coefficient

D: pipe diameter

$\overline{|u|}$: mean value of the velocity absolute value ($m.s^{-1}$)

Note that, in laminar regime, result $f\overline{|u|}$ is strictly constant.

By proceeding as for equations [2.13] and [2.14], we can eliminate u, obtaining:

$$\frac{\partial^2 P}{\partial t^2} + R\frac{\partial P}{\partial t} - C^2 \frac{\partial^2 P}{\partial x^2} = 0$$

This equation involves the damping term R.

2.8.5. *Resolution of the damping equation*

In any case, we have reached an equation with the form:

$$\frac{\partial^2 P}{\partial t^2} + D\frac{\partial P}{\partial t} - C^2 \frac{\partial^2 P}{\partial x^2} = 0$$

Let us verify if we can accept a solution of the form:

$$P = e^{\frac{-Dt}{2}} \cos\omega t \cos\gamma x$$

Indeed:

$$\frac{\partial P}{\partial t} = -\frac{DP}{2} - \omega \sin\omega t \cos\gamma x\, e^{-\frac{Dt}{2}}$$

$$\frac{\partial^2 P}{\partial t^2} = \frac{D^2 P}{4} - \omega^2 \cos\omega t \cos\gamma x\, e^{-\frac{Dt}{2}} + \omega D \sin\omega t \cos\gamma x\, e^{-\frac{Dt}{2}}$$

$$\frac{\partial^2 P}{\partial x^2} = -\gamma^2 \cos\omega t \cos\gamma x\, e^{-\frac{Dt}{2}}$$

By bringing these expressions into the equation, we have:

$$\left(C^2\gamma^2 - \omega^2 - \frac{D^2}{4}\right)\cos\omega t \cos\gamma x = 0$$

Consequently:

– We check whether a solution exponentially decreasing in time is possible.

– We note that pulsation ω and wavelength $1/\gamma$ are not arbitrary and are connected by:

$$C^2\gamma^2 - \omega^2 = \frac{D^2}{4}$$

– The boundary conditions allow us to determine γ.

2.8.6. *Weakening of a sound wave along a pipe for a single trip*

$$\frac{P}{P_0} = e^{-\frac{Rt}{2}}$$

$$\text{with} : R = \frac{f|\overline{u}|}{2D}$$

f: friction coefficient (f # 0.03)

D: pipe diameter (m).

If, for example:

$$|\overline{u}| = 2 \text{ m.s}^{-1} \qquad D = 0.1 \text{ m}$$

$$R = \frac{0.03 \times 2}{2 \times 0.1} = 0.3$$

For $P/P_0 = 1/3$:

$$t = \frac{2 \times 1.09}{0.3} = 7.3 \text{ s.}$$

If, in the pipe, the speed of sound is 1,500 m.s^{-1}, for the acoustic pressure to be divided by three, the distance covered will be:

$$L = Ct = 1500 \times 7.3 = 10950 \text{ m.}$$

This is a significant distance, yet is quickly reached by the many round-trips due to reflection at the pipe ends.

3

Block or Stop Valves and Control Valves

3.1. On valves in general

3.1.1. *Terminology*

Webster's dictionary provides the following definitions for valves:

Valve: a lid or cover so formed as to open a communication in one direction, and close it in the other.

Tap: a pipe to be inserted in a cask for drawing liquid, and stopped with a peg or spigot.

For present purposes, we will refer to a valve as any part of a pipe that enables:

1) Stopping or allowing the flow (in either direction) of a liquid. This covers both manual and automatic stop valves.

2) Restricting a flow stream by throttling, and modulating flow conveyed in a pipe. This covers control valves (which are automatic) and regulating valves (which are manual).

3.1.2. *The essential parts of a valve*

A valve comprises the following parts:

1) The body. For mass production, this can be obtained by forging, though in certain cases it requires direct casting (for moderate pressure) or even mechanical welding. The upper part of the body, if separate, is called the bonnet. Assembly of the body and the upper part is typically of the flange and seal variety, except at high pressures for which it is welded.

2) The shut-off system inside the body. This system can be of the following varieties: butterfly valve (rotating disk), knife (knife gate valves), membrane, etc. On closure, irrespective of its variety, the shut-off puts pressure on its seat. The principle for ball valves is, as we will see, quite different. In the special case of plug valve systems, the plug and its seat must have Brinell hardness over 400 if they have been stellated. If the plug sealing surfaces are Monel or stainless steel then the plug must have Brinell hardness of at least 15 above the seat value.

3) The shut-off control stem must, according to its variety, be able to:

– rotate: butterfly valves, ball valves, eccentric shut-off;

– slide: automatic control of valve gate, shutter, knife, plug valve, pinch valve, membrane valve;

– move in a helical movement: manual control of knife, gate, shutter, plug or membrane valve. The threaded stem is set in a fixed nut, but the shut-off should not turn and the stem should turn freely relative to this element.

If the operating temperature is above 300°C or below –45°C, an extension of the stem should be applied so that the control end can remain at a temperature between 0 and 70°C.

4) The control system acting on the stem can be:

– Manual: flywheel or lever. In both cases, the effort required by the maneuver should be limited:

Flywheel diameter or lever length (mm)	Effort (torque/diameter or torque/lever length: N)
$D \leq 150$	250
$150 < D \leq 350$	350
$D > 350$	400

– Automatic (typically used for remote control, consisting of a device that can be pneumatic (membrane or jack), hydraulic (jack) or electromagnetic (solenoid or front-rear motor). In block valves, the solenoid is faster than a pneumatic membrane. Pneumatic devices are mandatory in case of fire hazard (e.g. in the oil industry).

5) Pipe connections may be:

– flange, screw or welded (for DN less than or equal to 40);

– exclusively flanged or soldered (high pressure) for DN above 40.

3.1.3. *Sealing*

The seal is considered to be:

– average if leakage is in the order of a gas bubble (immersion test) or a drop of liquid;

– good, if leakage is zero.

Sealing is defined:

– as between the process fluid and the exterior atmosphere (body crossed by the stem);

– as between the upstream process fluid and the downstream process fluid (for the closed valve position). Upstream-downstream sealing by a stop valve must be good. That of a control valve may be only average.

There are three ways of maintaining sealing at the stem crossing with regard to the exterior:

– lining with single or double braid (with lantern piece to introduce coolant fluid). Double linings are used for pressures above 400 bars and for inflammable, toxic or simply leaking fluids (NH_3, H_2, CFC);

– with an O-ring for simple stem rotation;

– with a bellows for use on simple sliding stems, which like double lining are a requirement for high pressure and inflammable, toxic or leaking fluids. Bellows are more reliable than double linings and function satisfactorily after 10,000 cycles of pressure variation.

3.1.4. *Protection against corrosion and abrasion*

Elastomers are well known for their effective resistance to abrasion. Indeed, we will see that in certain valves, all surfaces in contact with the process fluid are made entirely of elastomers. This type of valve is also resistant to corrosion without requiring the use of an expensive alloy.

In general, valves cannot be used simultaneously at maximum pressure and at maximum temperature. The manufacturer's opinion is indispensable here.

3.1.5. *Protection against pressure and temperature*

Type of valve	P_{max} (bar)	T_{max} (°C)
Conventional plug valve	700	500
Gate	700	650
Lubricated ball valve	350	300
Non-lubricated ball valve	200	200
Butterfly	80	500
Membrane	15	130
Pinch	15	130

3.2. Different types of valves

3.2.1. *Valve categorization and study design*

We will examine the following valves:

– Plug valves	– Pinch
– Angle valves	– Butterfly
– Special plug valves	– Gate
– Eccentric shut-off	– Knife
– Ball	– Cage
– Membrane	– Multipath

The valve study design can be performed as follows:

1) description:
   - obturator type;
   - working mechanism;
2) action on fluid:
   - sealing: upstream-downstream;
   - sealing with respect to exterior atmosphere.
   - characteristic (see Figure 3.19);
3) mechanical and chemical protection:
   - pressure;
   - temperature;
   - corrosion;
   - abrasion;

4) price;

5) conditions for use:

– stop, regulation;

– types of fluid.

3.2.2. *Plug valves*

Here, shut-off occurs due to the plug lowering onto its seat. As we shall see, the plug varies according to the model. It is preferable for the upstream fluid to arrive under the plug, which eases plug opening. The stem connected to the plug operates in a sliding motion (typically vertical). This is particularly the case for control valves whose control comprises a circular membrane that changes shape to move the plug stem under the influence of compressed air. On the other hand, in stop valves together with manual control valves, stem movement is helical, and the stem is fixed in a threaded nut. Consequently, the maneuver is performed by means of a flywheel.

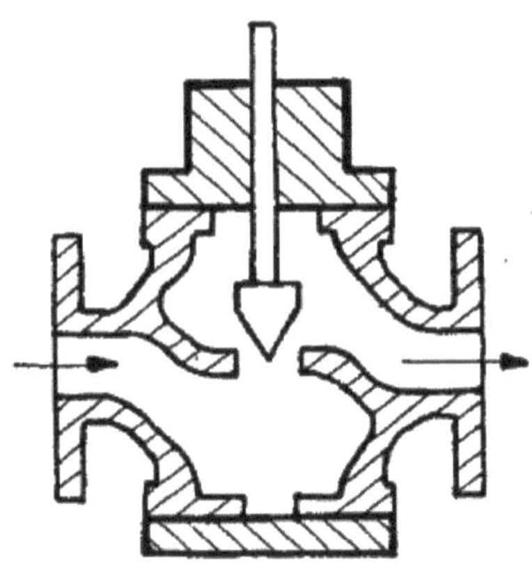

Figure 3.1. *Plug valve*

This type of valve is well sealed (upstream–downstream and to the exterior), which makes it advantageous for use blocking high pressure.

The presence of the valve produces a significant throttling in the process fluid, which, moreover, goes through a zigzag involving two changes in direction. Accordingly, there is a notable pressure drop across plug valves.

The characteristics of the valve can be exponential, linear or logarithmic (see section 3.3.14) according to the profile chosen for the plug (and seat). The best tuning results are achieved using a pointed plug. On the other hand, simple blocking may be achieved with a basic disc shape. In the latter case, the characteristic is logarithmic.

Plugs and their seat are always metallic, and consequently, effectively resist extreme temperatures. Moreover, the body is typically sufficiently resistant for plug valves to be used at high pressures. If noble alloys are used then corrosion protection is good.

On the other hand, the conveyed fluid must be clean, since if solid particles happen to stick between the valve plug and its seat, this can harm valve sealing and irreversibly damage both sides of the system.

The price of plug valves is significant, particularly for nominal diameters above 150 mm.

The above shows that plug valves should be used only for clean fluids, even for some corrosive fluids, irrespective of pressure or temperature limitations.

3.2.3. *Angle valves*

These are control valves used only for important relaxation. The relaxation ratio, that is, the ratio of upstream pressure to downstream pressure, can reach 20.

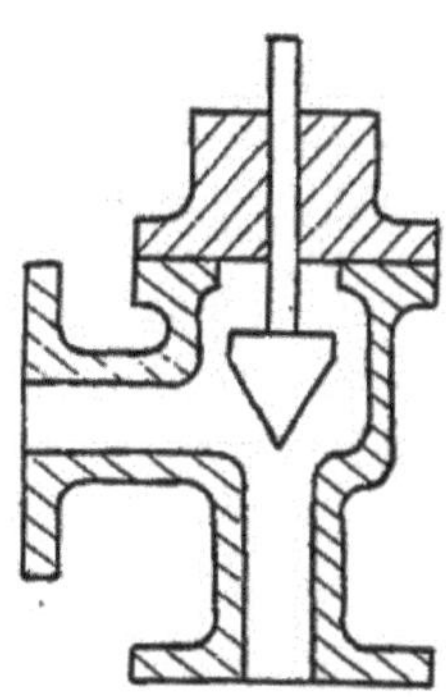

Figure 3.2. *Angle valve*

Angle valves have the advantage of producing little noise. Indeed, fluid vaporization by valve relaxation produces fluid droplets dispersed at high velocity by the vapor and, as fluid direction changes, the liquid hits the pipe wall, which generates noise.

The fluid path in an angle valve is less sinuous than that in a conventional valve, with fewer fluid impacts and a less noisy valve.

Furthermore, if the fluid contains suspended solids, angle valves avoid not only producing noise, but also the erosion of the valve body by the solid. Nonetheless, we must remember that only direct or near-direct pathway valves are typically used for suspended solids as they avoid blockages and do not contain valve plugs and seats which could promote erosion.

3.2.4. *Other plug valves*

Where the plug is sloped, this results in *unfolding* the fluid zig-zag path and thereby attenuating the changes in the direction typically observed, reducing the pressure drop.

For high upstream pressures, there are plug valves comprised of two valves on the same stem. The upstream fluid arrives between two plugs, which neutralized the pressure efforts at work on the plugs (and the stem). The opening and closing mechanisms are eased, and any regulatory precision required is obtained.

Figure 3.3. *Balanced plug valve*

When solid deposit (or crust) is suspected to block a draining pipe positioned on the bottom of a tank, we fit a breaking crust valve. On opening, the plug lifts, breaking any deposit that has formed. The vertical stem that activates the plug is accessed under the tank, which is an elbow-shaped valve body.

3.2.5. *Eccentric shut-off control valve*

When a regulation valve is only slightly open (less than 30% of the seat cross-section, for example), the upstream–downstream pressure difference causes a parasitic effect on the plug, disturbing valve calibration.

To solve this problem, Masoneilan developed valves with a spherical-cap shut-off that acts against a circular orifice, obstructing it to the desired amount.

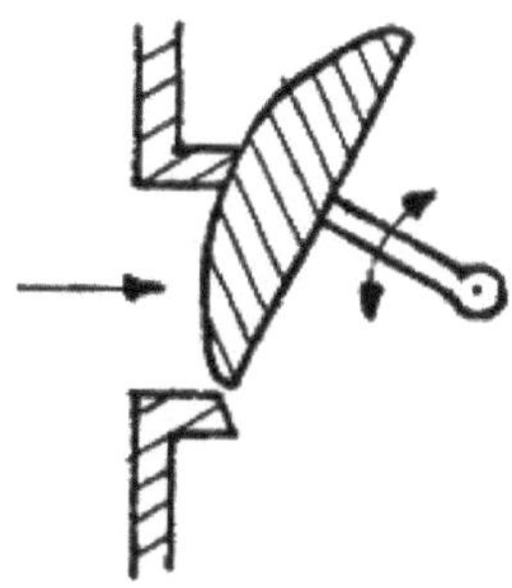

Figure 3.4. *Valve with eccentric shut-off*

Using this device, calibration remains precise irrespective of the opening. Its operating range (O.R.) approaches 100%, while typically it is only 100 – 30 = 70% for simple plug valves.

Eccentric shut-off control valves were conceived for pressures below 100 bars and for nominal diameters of no more than 300 mm.

In process terms, eccentric shut-off control valves have the same use conditions as plug valves.

3.2.6. *Ball valve*

A cylindrical, conical or spherical metallic part with a wide orifice traversing it from one side to the other. This part, pivots on its axis symmetrically in a seat set in the valve body.

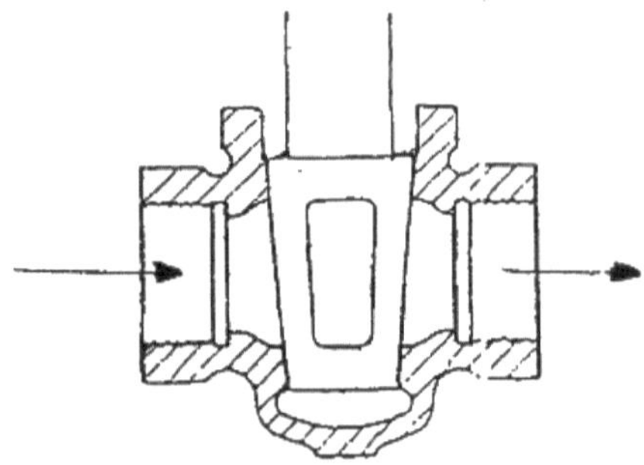

Figure 3.5. *Conical plug cock valve*

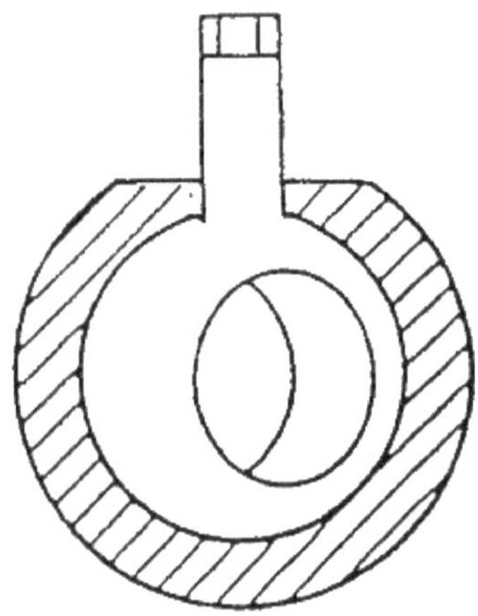

Figure 3.6. *Spherical ball valve*

When the orifice is turned toward the pipe, the valve is in the wide open position. For complete closure, the ball must be turned a quarter turn.

This maneuver is performed using a lever that can be operated manually or by means of a hydraulic or pneumatic actuator, or by a solenoid.

The sealing surface of the plug is typically metallic for both cylindrical and conical ball valves. Consequently, conical ball valves can exhibit a tendency to jam if not lubricated. However, it is always possible to line the sealing surface with an elastomer or PTFE. Indeed, this is always the case for spherical balls. In order to lubricate conical balls, we introduce oil by means of groves along the cone.

The lining or lubrication ensures a moderate value for the couple as required for the rotation movement and also ensures optimal sealing, even for gases.

The orifice cross-section can be circular and of an equal diameter to that of the pipe. In such cases, the valve is a *direct pathway* and its pressure drop is practically zero. This valve is useful as a control valve, presenting sigmoidal characteristics (see section 3.3.14). Indeed, with the valve half open, the variation in the flow cross-section is maximum for a given variation in the floating ball rotation angle. If we require a different characteristic, the orifice cross-section would no longer be circular and would be less than that of the pipe. The valve would then be a *quasi-direct pathway*.

Plug cock and ball valves are only resistant to moderate temperatures and pressures. Their size is typically limited to a nominal diameter of 300 mm.

Significant variations in fluid temperature can provoke differential dilatations resulting in blockages and a loss of sealing. Consequently, we bypass the valve with a pipe of very low diameter to facilitate opening the closed valve. Opening this bypass balances the temperatures upstream and downstream of the valve.

Ball valves are relatively resistant to certain corrosions if the plug cock or ball metal is suitable and if the seat is lined. For simple use, the plug cock or ball itself is made of PTFE.

Direct pathway valves are recommended for slurries. The significant size of the orifice itself makes blocking up improbable, and even if this does occur, flushing with a clean liquid through the open valve should be enough to remove the obstruction. A well-lubricated sealing surface should be selected. Indeed, if a solid particle gets between the plug cock or ball and its seat, and if the sealing surfaces are both metallic, the particle could provoke a jam and scratch the surfaces, damaging the sealing.

Ball seals are moderately priced. If the valve parts are made of standard metal, the price can be very reasonable.

These valves are typically used for blocking and, occasionally for control (particularly for slurry).

3.2.7. *Membrane valve*

A stem ending with a button presses the membrane against the diaphragm, which provokes closure. The diaphragm is in the form of a crescent moon and both partially and continuously closes the passage of fluid.

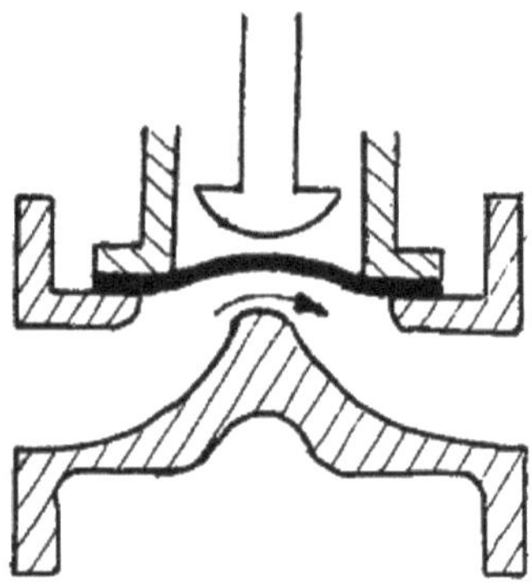

Figure 3.7. *Membrane valve*

The operation mechanism is the same as that of a plug valve but, here, the sealing with respect to the outside is perfect as the stem does not enter the volume occupied by the process fluid. Nonetheless, this disposition presents the inconvenience of being unsuitable for use in a vacuum, as the membrane would apply itself to the diaphragm, meaning that the valve would remain closed. Inversely, the process fluid pressure must not exceed a value in the order of 15 barG in order to avoid membrane rupture.

Upstream–downstream sealing is good and the drop in pressure is lower than that of a plug valve (though by no means negligible). When used for control purposes, membrane valves offer an unchanging characteristic approaching the logarithmic model.

The operating temperature is limited by the presence of elastomer in the membrane.

If the body is made of a noble alloy, ebonite or doubled with an elastomer, this valve presents excellent corrosion protection. Furthermore, its resistance to abrasion is high.

The membrane valve is recommended for fast blocking due to the cushion effect of the membrane that presses on the diaphragm attenuating the inertial impact (liquid hammer) of the pipe fluid. Occasionally, this valve is used for control, specifically for slurry. It is a requirement for toxic or leaking fluids due to its good outside sealing.

3.2.8. *Pinch valve*

An elastomer sleeve is pinched between two bars fixed in place by a clamp. The upper pinch bar is lowered by a stem whose movement is helicoidal in manual operation and sliding in automatic operation. It is also possible to pinch the sleeve using compressed air. This valve cannot be used in vacuum as it would remain closed.

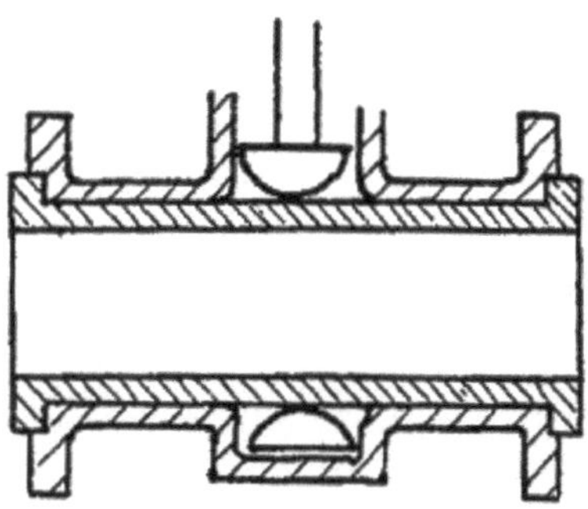

Figure 3.8. *Pinch valve*

The pinch valve is perfectly sealed to the exterior and has good upstream–downstream sealing. Similar to the membrane valve, it can only withstand limited pressure in the process fluid.

This is a direct pathway valve with very low pressure drop. It is used as control valve with a flow characteristic approaching the logarithmic shape (see section 3.3.14).

Owing to the presence of elastomer in the sleeve, protection against extreme temperatures is limited.

The pinch valve is highly resistant against corrosion and abrasion, and furthermore, it is inexpensive.

While used as a control valve, its main use is as a block valve. It is recommended for corrosive fluids, slurries and even for turbid slurry.

3.2.9. *Butterfly valve*

Here, a disk turns on the diameter at a cross-section and can thus allow or stop the flow of fluid. In closed position, it is perpendicular to the process flow. In open position, the obstruction caused is low.

Between opening and closing, the disk has to take only a quarter turn. The operation can be performed manually using a lever. It can be performed automatically by means of an actuator. During the opening process, high torque may be required.

Exterior sealing requires careful study of the trunnions that allow for disk rotation, while upstream-downstream sealing also requires a seal of sophisticated conception (fixed on the body) on which the closing disk comes to rest.

When completely open, the pathway open to the fluid is only near-total due to the presence of the rotation axle and, more simply, the disk thickness. Nonetheless, the pressure drop remains low. The flow characteristic is slightly sigmoidal (parabolic for $\theta < 50°$). Here, the characteristic is the variation of the flow pathway cross-section according to the rotation angle.

As differential dilatations are still possible, the use of this type of valve is limited in temperature. It is also limited in terms of pressure (sealing *vis-à-vis* exterior).

The valve body and the disk can be readily coated (in PTFE for example) and the corrosion resistance is excellent as a consequence. On the other hand, the butterfly valve may only be used with clean fluids, as if a solid particle should be trapped, it could prevent complete valve closure.

Butterfly valves are moderately priced, which explains their use at diameters in the order of 6 m in hydraulic plants.

It is used as both a control valve and a block valve.

3.2.10. *Gate valve*

In these valves, closure is performed by lowering a transversal wall into the process flow, with the shut-off resembling a sliding door.

The gate can be:

– Parallel side, with the seal provided by two O-rings, this can also be referred to as a spade valve.

– Wedge shaped, which allows for a better seal; here, the opposing faces are metallic. These are wedge-type valves.

The shut-off is inside the valve body, which consequently includes a bulge where it is lodged when lifted.

The gate is operated using a stem whose movement can be helicoidal (manual operation using flywheel) or sliding (automatic operation).

The seal is just as good as that of plug valves, with respect to both the outside and upstream–downstream flow.

As this is a direct pathway valve, the pressure drop is very low. These valves are not used as control valves due to violent eddies (generating vibration) appearing downstream if the valve is partially closed. As a consequence, the notion of flow characteristic is of no interest here. Fluid can flow in both directions.

These valves present effective protection against pressure and moderate temperatures. They are not particularly well suited for abrasive suspensions and only offer corrosion resistance if an appropriate metal alloy is selected.

In reality, however, as these valves are widely used as standard models for blocking operating fluids (water, low-pressure steam, compressed air), they are inexpensive.

3.2.11. *Knife valve*

This is a special type of gate valve. The back edge of the wedge (the *knife edge*) often crosses the valve body, which requires particular sealing.

What distinguishes the knife valve from the shut-off of a classic gate valve is the cutting capability of the *knife edge*:

– its thickness is low;

– it is cut into a bevel on its cutting edge.

Thus, on closure, it is able to cut any fiber present or to clear a passage in a divided solid flowing in the conduit.

Owing to this capacity, the knife valve is only used for blocking, particularly:

– in the paper industry (where they treat suspensions of fibers);

– in conduits in which solid flows (closure of silo exit).

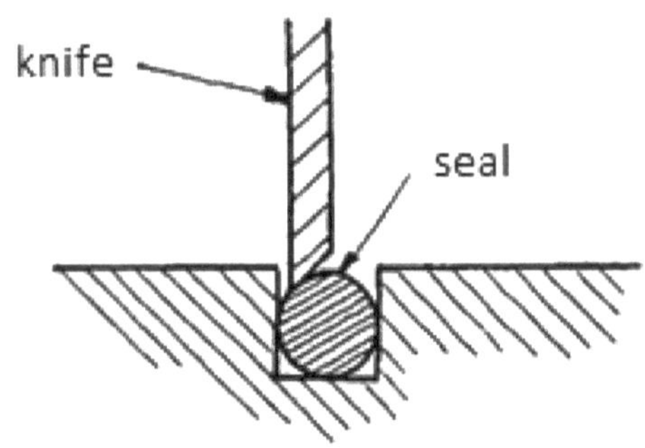

Figure 3.9. *Knife valve*

3.2.12. *Cage valve*

This valve comprises a piston sliding in a cylinder (the cage) pierced with holes. The piston height is close to that of the cage so that when enclosed by the cage the holes are entirely blocked and the valve is closed. Typically, the fluid flows from inside of the cylinder to the outside (though not always). The piston guidance by the cage ensures good protection against vibration.

As the piston is equipped with rings, upstream–downstream sealing is excellent, even for high upstream pressures.

The pressure drop depends on the number of holes pierced in the cage. There are two ways of obtaining a fixed form of flow characteristic for a valve beforehand:

– by modifying the number of holes located on the circumference, as we move on the cylinder generatrix, though this is not always an easy economical option;

– by using a cam positioner.

The natural flow characteristic of a cage valve is linear, which corresponds to a uniform spread of holes in the cage (see section 3.3.14).

The operating range (O.R.) is of 100% due to calibration precision not depending on the piston position.

Cage valves are effective at high temperatures and pressures and, when made in noble alloys, are resistant to corrosion. The price of these valves is high. They are only to be used for clean fluids.

In reality, their main role is in allowing for significant relaxation with minimum vibration, noise and, with liquids, cavitation. They are often compared with angle plug valves used for high differential values of relaxation pressure. Indeed, by installing several coaxial cages fitted with staggered holes from one cage to the next, we obtain an offset arrangement equivalent to the muffler of a car. Furthermore, with this device, the relaxation is progressive and the pressure does not drop in such a way as would lead to cavitation phenomena.

Cage valves are control valves that, according to their size, can be used for flows between 1 and 100 $m^3.h^{-1}$.

3.2.13. *Multipath valves*

1) Four-way block ball valves.

Figure 3.10 depicts the division obtained for a gas between two columns of drying gas, of which one is process gas and the other is regeneration gas.

2) Three-way valves.

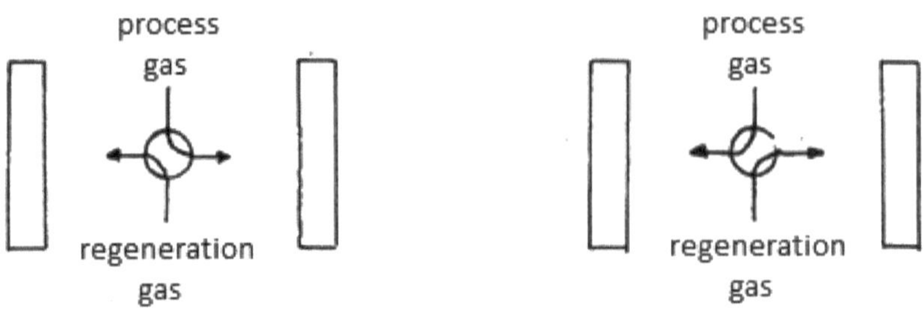

Figure 3.10. *Four-way block ball valve*

Three-way valves allow for flow from A to be directed toward B or toward C.

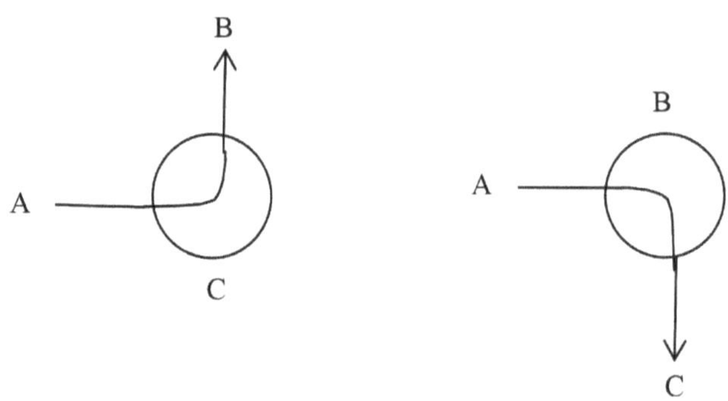

Figure 3.11. *Three-way valves*

3) Two-way drawer valve for thick slurry.

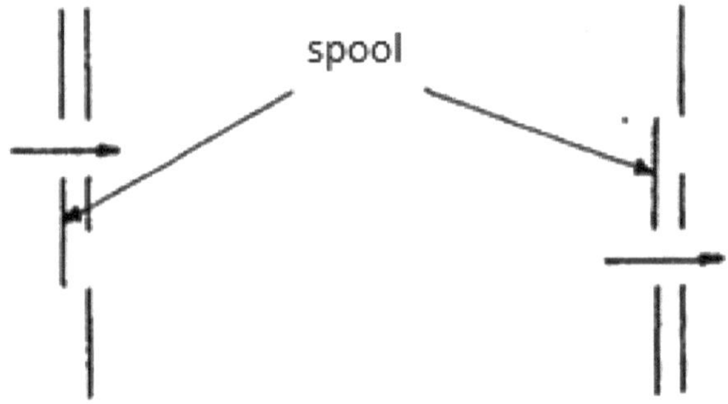

Figure 3.12. *Principle of a two-way drawer valve*

4) Three-way drawer valve for thick slurry.

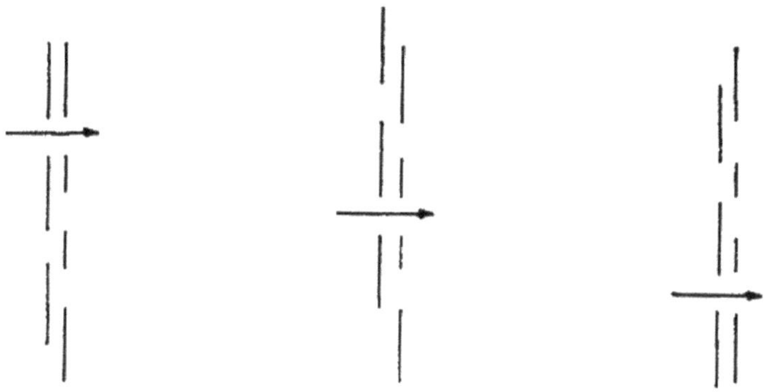

Figure 3.13. *Principle of a three-way rotary spool*

The tap system for slurry is actioned by hydraulic systems.

5) Three-way plug control valve for clean fluid.

Figure 3.14. *Principle of a three-way plug block valve*

In developing three-way plug valves, one key aspect must be respected: the fluid must always flow in such a way that the upstream pressure acts under the plug.

Figure 3.15. *Principle of a three-way plug mixing valve*

3.3. Control valve choice and calculation

3.3.1. *Purpose of control valves*

These valves allow for the fluid or gas flow to be modulated by throttling the process flow.

The flow needs to be modulated accordingly when we encounter the following situations:

– We need to keep a constant level of fluid in the upstream device or in the downstream device.

– We need to provide exactly the right level of heat required to keep the contents of a device at a constant temperature or pressure. Heat (and cold) are typically transported by a utility fluid (steam, cold water, brine, ethylene glycol solutions), though this is not obligatory and a process fluid can suffice.

– We need to alter the flow of an acid or alkaline in order to maintain the pH of the fluid held in the downstream device.

– We need to regulate the flow of a reagent proportionally to the flow of another reagent.

– We need to maintain a constant flow independently of external conditions. The value of the flow setpoint must be selected within an O.R.

This list is not exhaustive. In the previous cases, as well as in the last, the valve must allow for changes in the flow within an O.R. set below a minimum flow Q_m and above a maximum flow Q_M, independently of disturbances due to external conditions.

To borrow the terms used by electricians, the valve is a variable resistor that must be able:

– to increase in order to compensate for a reduction in resistance of the line or an increase in the *potential* between the line ends;

– to reduce in order to compensate for an increase in the resistance of the line or a reduction in the *potential* difference existing between the line ends.

3.3.2. *Flow in a valve*

As we have seen, the purpose of a control valve is to throttle the process flow. Inside the valve, we find a device (for example, valve plug, more or less withdrawn from its seat) in which the cross-section pathway for the fluid is minimum. Consequently, the velocity is maximum at this point, and, according to Bernoulli's theorem, the pressure is minimum. This is expressed by the curves in Figure 3.16.

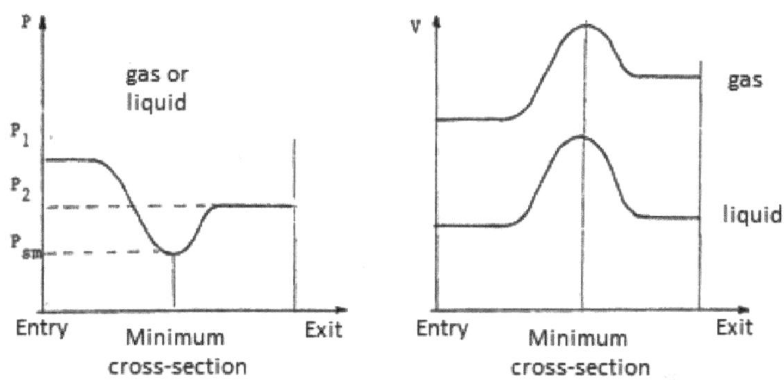

Figure 3.16. *Development of velocity and pressure in a valve*

Pressure P_2 on exit is inferior to pressure P_1 on entry due to the singular drop in pressure created by the throttling device. Unlike with fluids, gases would have a higher velocity on exit than on entry since pressure would be lower on outlet.

For a liquid, ebullition is possible at pressure P_{sm}. The bubbles would then condense abruptly when the pressure increases.

3.3.3. *Thermodynamic approach and calculation principle*

The process of crossing a valve (and control valves in particular) is always a free relaxation without any work production.

Only thermodynamic calculations allow us to obtain the exit temperatures of gases according to the downstream pressure and upstream conditions (pressure and temperature).

We have seen the selection criteria for valves in general. We will also look at how to obtain the flow and drop in pressure, allowing us to correctly scale a valve, its *valve coefficient*, which is the C_V of the valve in question.

3.3.4. *Calculation of C_V with flow in volume*

For a fluid, the unit C_V corresponds to a flow of water at 60 °F equal to 1 $gal.min^{-1}$ under a differential pressure ΔP equal to 1 pound per square inch (psi). With these units:

$$C_V = Q\sqrt{\frac{d}{\Delta P}}$$

d is the density of liquid in the suction conditions taken relative to water at 60 °F, that is, 15.5°C. The correspondence with metric units is as follows:

$$\frac{1\ m^3}{h} = 4.4028\ \text{gpm}$$

$$1\ \text{bar} = 14.504\ \text{psi}$$

Hence, for a liquid:

$$C_V = \frac{4.4028}{\sqrt{14.504}} Q\sqrt{\frac{d}{\Delta P}} = 1.156 Q\sqrt{\frac{d}{\Delta P}}$$

For a gas, the definition is the same as that of density relative to water that intervenes. This density is thus expressed as:

$$d_G = \frac{M}{22.4} \times \frac{273}{T} \times \frac{P_1}{1.013} \times \frac{1}{999} \times \frac{1}{Z_1}$$

M is the molar mass of the gas.

The compressibility coefficient Z_1 (in the suction conditions) intervenes as:

$$PV = Z\,n\,RT \quad \text{and} \quad \rho_G = \frac{n\,M}{V} = \frac{P}{ZRT}\left(kg.m^{-3}\right)$$

There are curves that provide the compressibility coefficient according to the reduced pressure P_1/P_c and the reduced temperature T_1/T_c.

Furthermore, in the definition of C_V, the real flow Q_R can be obtained by correcting flowrate Q_N taken in normal conditions, that is, 0°C and atmospheric pressure.

$$Q_R = Q_N \times \frac{1.013}{P_1} \times \frac{T_1}{273} \times Z_1$$

Finally:

$$C_V = 1.156 Q_N \sqrt{\frac{M}{22.4} \times \frac{T_1}{273} \times \frac{1.013}{P_1} \times Z_1 \times \frac{1}{999} \times \frac{1}{\Delta P}}$$

That is, for a gas:

$$C_V = \frac{Q_N}{2125}\sqrt{\frac{MT_1Z_1}{P_1\Delta P}}$$

3.3.5. *The K_V in the international system of units*

1) Gases

By definition:

$$K_V = Q_N^* \sqrt{\frac{M}{22.42} \times \frac{T_1}{273} \times \frac{1.013.10^5}{P_1^*} \times Z_1 \times \frac{1}{999} \times \frac{1}{\Delta P^*}}$$

In this expression, Q_N^* expresses flow in $m^3.s^{-1}$ with P_1^* and ΔP^* expressed in Pa.

Yet:

$$\sqrt{\frac{1.013}{22.42 \times 273 \times 999}} = \frac{1}{2457}$$

Hence:

$$K_V = \frac{Q_N^*}{2457}\sqrt{10^5}\sqrt{\frac{MT_1Z_1}{P_1^*\Delta P^*}} \qquad [3.1]$$

On the other hand, C_V can be written as:

$$C_V = \frac{1.156}{2457} Q_N \sqrt{\frac{MT_1Z_1}{P_1 \Delta P}} \qquad [3.2]$$

In this last expression, Q_N expresses flow in $m^3.h^{-1}$ with P_1 and ΔP expressed in bars.

However:

$$1\ m^3.s^{-1} = 3600\ m^3.h^{-1}$$

$$1\ bar = 10^5\ Pa$$

Hence, for numbers expressing the quantities:

$$\frac{Q_N}{Q_N^*} = 3600 \quad \text{and} \quad \frac{P}{P^*} = \frac{1}{10^5}$$

By dividing equation [3.2] by equation [3.1] expression by expression, we obtain:

$$\frac{C_V}{K_V} = \frac{1.156 \times 10^5 \times 3600}{\sqrt{10^5}} = 1.316.10^6$$

2) Liquids

By definition:

$$K_V = Q^* \sqrt{\frac{d}{\Delta P^*}} \quad \text{and} \quad C_V = 1.156\, Q \sqrt{\frac{d}{\Delta P}}$$

Hence:

$$\frac{C_V}{K_V} = 1.156\sqrt{10^5} \times 3600 = 1.316.10^6$$

3) Conclusion

Whether it is a gas or a liquid, the ratio C_V/K_V is clearly the same, since C_V and K_V are properties of the valve and not of the processed fluid. Finally, we obtain the simple relationship:

$$K_V = 0.7599.10^{-6}\, C_V$$

With (leaving the asterisks),

$$\text{Liquids:}\ K_V = Q\sqrt{\frac{d}{\Delta P}}$$

$$\text{Gases:}\quad K_{V0} = 0.129 Q_N \sqrt{\frac{M T_1 Z_1}{P_1 \Delta P}}$$

(effectively: $\sqrt{10^5}/2457 = 0.129$).

3.3.6. *Calculation of C_V with flow in mass*

For liquids, it suffices to replace Q by:

$$Q = W/\rho_L = W/(d \times 999) \qquad (W \text{ in kg.h}^{-1})$$

Hence:

$$C_V = \frac{1.156\ W}{999\sqrt{d\Delta P}} = 1.157.10^{-3}\frac{W}{\sqrt{d\Delta P}}$$

For gases, we replace Q_N with:

$$Q_N = W/\rho_{GN} = W/\left(M/22.4\ Z_0\right)$$

Z_0 is the compressibility coefficient of gas in normal conditions. Typically, the value Z_0 is very close to 1.

Hence:

$$C_V = W \times \frac{22.4}{2125}\sqrt{\frac{T_1 Z_1}{P_1 \Delta P M}} = \frac{W}{94.8}\sqrt{\frac{T_1 Z_1}{P_1 \Delta P M}}$$

NOTE.–

Let us recall the metric units that we have selected as the most widespread:

ΔP: bars

T_1: Kelvin

M: $kg.kmol^{-1}$

Q: $m^3.h^{-1}$

W: $kg.h^{-1}$

3.3.7. *Laminar flow of a liquid*

Only a liquid, as opposed to a gas, can cross a control valve in laminar regime.

Schlumberger uses the following formula:

$$C_V = \frac{1}{K_L}\left(\frac{Q\mu}{\Delta P}\right)^{2/3}$$

With:

$$25 < K_L < 35 \text{ according to the valve type}$$

μ is the viscosity of the liquid in centipoises.

3.3.8. *The meaning of C_V*

The pressure drop in the valve arises in large part due to the restriction organ (plug, for instance), so that, between the entry and this organ, Bernoulli's equation can be employed in all its simplicity. We can characterize the minimum cross-section (corresponding with the restriction) by the indices sm:

$$P_1 - P_{sm} = \frac{\rho V_{sm}^2}{2} - \frac{\rho V_{entrée}^2}{2} \# \frac{\rho V_{sm}^2}{2} = \frac{W^2}{2\rho A_{sm}^2}$$

A_{sm} is the minimum cross-section. Furthermore:

$$V_{sm} >> V_{entry}$$

The pressure drop across the valve is the fraction p (loss) of the difference P_1–P_{sm}. This fraction is not recuperated in the divergence connecting the restriction to the valve exit. Therefore:

$$\Delta P = p\left(P_1 - P_{sm}\right) = \frac{pW^2}{2\rho A_{sm}^2} \qquad [3.3]$$

Indeed, by definition of the C_V,

$$W = \frac{999}{1.157} C_V \sqrt{d\Delta P}$$

However:

$$1 \text{ kg.s}^{-1} = 3{,}600 \text{ kg.h}^{-1}$$

$$1 \text{ Pa} = 10^{-5} \text{ bar}$$

$$d = \rho/999$$

It follows that we can replace the traditional metric units with those of the international system. This results in:

$$W = \frac{C_V}{3.6 \times 1.157\sqrt{999 \times 2 \times 10^5}}\sqrt{2\rho\Delta P}$$

That is:

$$W = 0.17.10^{-4} C_V \sqrt{2\rho\Delta P} \qquad [3.4]$$

However, after formula [3.3]:

$$W = \frac{A_{sm}}{\sqrt{P}}\sqrt{2\rho\Delta P}$$

Hence:

$$0.17.10^{-4} C_V = A_{sm}/\sqrt{P}$$

3.3.9. *Cavitation of a fluid in a liquid*

Cavitation appears when pressure P_{sm} at the minimum cross-section (that is, the minimum pressure in the valve) is below the vapor pressure π of the liquid. In these conditions:

$$P_1 - P_{sm} \geq P_1 - \pi$$

Or again:

$$\frac{\Delta P}{p} \geq P_1 - \pi$$

And finally:

$$\Delta P \geq p(P_1 - \pi)$$

In reality, this is not true cavitation but rather vaporization followed by abrupt condensation.

3.3.10. *Limiting flowrate of a liquid entering at its boiling point*

Liquids partially vaporize due to the pressure drop on crossing the valve. We will show that the presence of the resulting vapor limits the flowrate in terms of the practical values of the differential pressure ΔP.

We should begin by noting that the pressure drop due to the acceleration of the vapor formed is low, even negligible relative to the total pressure drop. The dominant factor is the high velocity of the liquid–vapor mixture.

To the nearest ½ factor, the above is equally applicable for the kinetic dispersion energy on leaving the valve, so that we can accept that relaxation is isenthalpic.

According to this hypothesis, and with a good approximation, the vaporized fraction (in mass) is:

$$X = C_L \Delta T / L_1$$

L_1 is the latent vaporization heat at the entry temperature T_1 ($J.kg^{-1}$).

Let us call T the boiling temperature of the liquid at pressure π and develop T to the first order relative to π:

$$T_2 = T_1 + (\partial T / \partial \pi) \Delta \pi + O (\Delta \pi)^2$$

Since:

$$\Delta P = \Delta \Pi = P_1 - P_2, \text{ comes to:}$$

$$T_2 - T_1 = \Delta T = (\partial T / \partial \pi) \Delta \pi$$

And:

$$X = \frac{C_L}{L_1}(\partial T / \partial \pi)\Delta P$$

Or even:

$$\Delta P(X) = \frac{X}{\frac{C_L}{L_1}.\frac{\partial T}{\partial \pi}}$$

By accepting the validity of the Lockhart–Martinelli parameter for two-phase flow, we can write:

$$\Delta P(X) = \phi^2(X)\Delta P_L$$

ΔP_L is the pressure drop of liquid *without* vaporization crossing the same valve and at flow W equal to the total entry flow.

Yet ϕ is a function of X, that is, of $\Delta P(X)$ and can be written as:

$$\Delta P_L(X) = \Delta P(X) / \phi^2\{\Delta P(X)\}$$

Figure 3.17 shows the variations in ΔP_L (X) according to X.

We note that ΔP_L quickly increases to a maximum and then remains near constant and equal to ΔP^* when X varies from 0 to 0.25. Yet, this interval covers nearly all relaxations with vaporization occurring through a valve. The constancy of ΔP_L results in that of the total flow W when X (that is, ΔP) increases. Note that ΔP can only increase from zero to P_1.

Thus, if there is any vaporization, it can be written, approximately, as:

$$C_V = \frac{1.157.10^{-3} W_{lim}}{\sqrt{d\Delta P^*}}$$

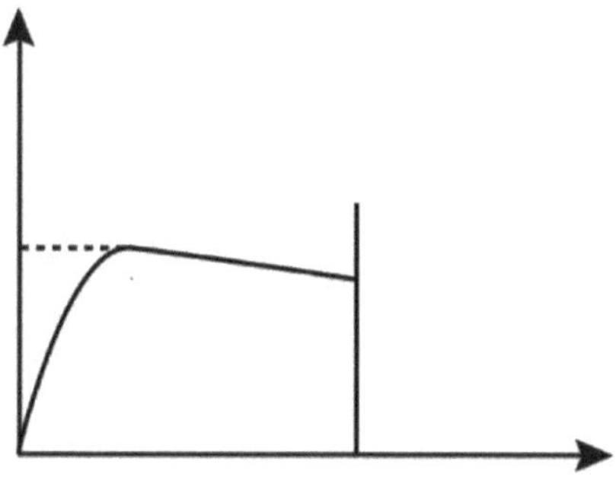

Figure 3.17. *Variation in ΔP_L according to the fraction vaporized*

ΔP^* is practically a function of P_1 and the critical pressure P_c of the liquid. This critical pressure sets the value of $\partial T / \partial \pi$. Hence:

$$\Delta P^* = C^2 \left[P_1 - \left(0.96 - 0.28 \sqrt{\frac{\pi}{P_c}} \right) \pi \right]$$

Coefficient C is assessed differently according to the manufacturers, which requires special consultation.

ΔP^* is in fact the pressure drop at the start of vaporization ΔP_{VA}:

$$\Delta P_{VA} = \Delta P^*$$

3.3.11. *Conclusion for liquids*

For a given valve, variations in flow W according to ΔP are represented in Figure 3.18.

ΔP_{CA} and ΔP_{VA} are, respectively, the pressure drops corresponding to the start of cavitation and the start of vaporization.

3.3.12. *Relaxation coefficient for gases*

The expressions that we have seen for C_V suppose that the gas behaves in the same way as a liquid. In reality, the density of a gas varies over its course through the valve.

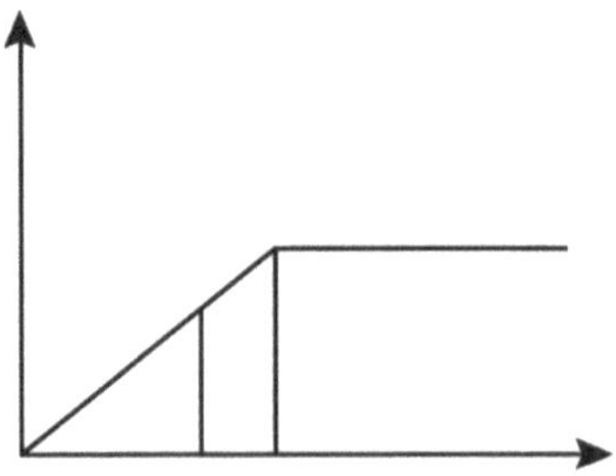

Figure 3.18. *Flow variation according to pressure drop*

In fluid mechanics, we show that for an isentropic variation parameter in a convergent–divergent, the ratio of the real gas flow to the supposed liquid flow is a complicated function of γ (that is, C_P/C_V), of P_{sm}/P_1 (P_{sm} is the pressure at the restriction) and of relationship m of the restriction cross-sections and the valve entry (see Brun *et al.* tome I [BRU 68, p. 375]).

Schlumberger proposes a simplified relationship for Y, posing:

$$Y = \frac{\text{real flow in mass}}{\text{flow of supposed liquid}}$$

3.3.13. *Sonic regime for gases*

This regime appears when the speed of sound is reached by the gas at the place of minimum cross-section (that is, the place where the velocity is maximum) and we see that (see Brun *et al.* tome I [BRU 68, p. 389]):

$$\frac{P_m}{P_1} = \left[\frac{2}{\gamma+1}\right]^{\frac{\gamma}{\gamma-1}} = \left[1 - \frac{\gamma-1}{\gamma+1}\right]^{\frac{\gamma}{\gamma-1}} \# 1 - \frac{\gamma}{\gamma+1}$$

Therefore:

$$\frac{P_1 - P_m}{P_1} \# \frac{\gamma}{\gamma+1}$$

If r is the recovery coefficient:

$$P_2 - P_m = r(P_1 - P_m)$$

By eliminating P_m between these two equations, we reach:

$$\frac{P_1 - P_2}{P_1} = \frac{\Delta P_S}{P_1} \# \frac{\gamma(1-r)}{1+\gamma} = \frac{\gamma p}{1+\gamma}$$

If $\Delta P > \Delta P_S$, the speed of sound does not increase significantly in the restriction, but the density of the gas flowing is proportional to P_1. The flows W and Q_N are then also proportional to P_1. By replacing ΔP with ΔP_S, we reach the values of C_V.

3.3.14. *Flow characteristic of a control valve*

In principle, a valve *characteristic* refers to the curve that represents the flow variations according to a mechanical parameter:

– either the lateral displacement of the stem (or, more precisely of the relative displacement X/X_∞ or X_∞ corresponding to the total opening);

– or θ the rotation angle of the stem.

In practice, without losing any precision, we can substitute the available cross-section of the fluid pathway for the fluid flowrate. Indeed, this flow is proportional to the passage cross-section. Ultimately, the characteristic announces variations in the percentage of the A/A_∞ ratio according to one of the two mechanical parameters above, in which A is the cross-section corresponding to a given opening, and A_∞ is the cross-section corresponding to the wide open valve:

$$100 \times A/A_\infty = f(X/X_\infty) \text{ or } 100 \times A/A_\infty = f(\theta)$$

Note that θ always varies from 0° to 90°.

We can primarily distinguish three ideal types of characteristics:

– linear

– exponential

– logarithmic.

We can add a fourth type to the three above; though it does not have any ideal character, it is derived from the practical configuration of certain valves. This is the sigmoidal type.

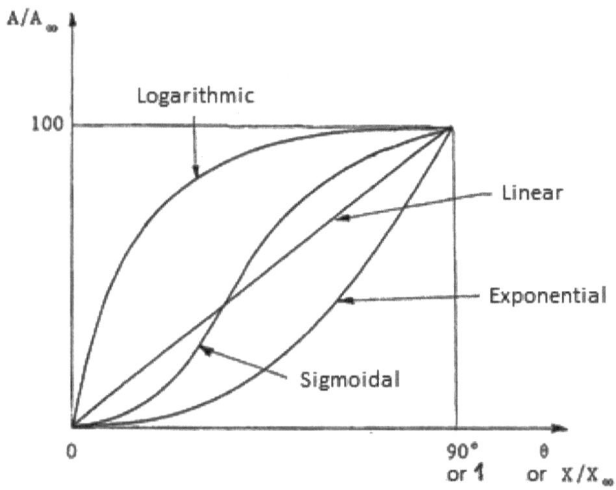

Figure 3.19. *Characteristics of regulation valves*

The choice of these three theoretical characteristics is made according to the following criteria:

1) Linear characteristic:

This characteristic is suitable when the required O.R. is limited or if the control loop is stable (the process gain is positive and varies little). This is the case for flow and level control in particular. In nominal flowrate, the valve is open at approximately 60%.

2) Exponential characteristic:

This is recommended if the O.R. needed is extensive and if the pressure drop is low (below 30%) relative to that of the line. This characteristic offsets a decreasing gain with the opening as is the case for control valves for temperature and pressure. At nominal flow, the valve is open at approximately 70%.

3) Logarithmic characteristic:

This is seldom used, except for fast-acting block valves.

4) Sigmoidal characteristic:

This is an imperfect substitute for the linear characteristic.

3.3.15. *Operating range*

1) Required O.R. (required O.R.):

Typically, an installation is set up for a given nominal flow Q_N. However, often a certain oversizing S_u is allowed for, which corresponds to a maximum flow Q_M:

$$Q_M = S_u \times Q_N$$

Furthermore, random disturbances or the need to operate the installation with a reduced regime requires the minimum flow Q_m which is deduced from the nominal flow by multiplication with the flexibility coefficient S_o less than 1:

$$Q_m = S_o \times Q_N$$

Understanding of these two flows establishes the O.R. required as follows:

$$\text{required O.R.} = \frac{Q_M - Q_m}{Q_m} \times 100$$

2) Available O.R. (O.R. available):

The total range of theoretical variations possible for valve opening is 100%. However, approaching closure, some valves present instability due to pressure on both sides of the closure device.

In other words, the total range is the sum total of the imprecision range (I.R.) and the available O.R. (O.R. available for satisfactory precision)

$$100 = \text{I.R.} + \text{O.R. availble}$$

Thus, for some plug valves, the I.R. is equal to 30%, resulting in an available O.R. of:

$$\text{available O.R.} = 100 - 30 = 70\%$$

3) Conclusion:

For satisfactory operation, we must have:

$$\text{available O.R} > \text{required O.R.}$$

NOTE.–

The drawback of oversizing a valve is clear. Indeed, if, over the same range, the O.R. is constant, oversizing can lead to an increase in the value of the minimum flow below which flow is imperfect. As a consequence, minimum flow is potentially greater than the flow Q_m required by the installation for low flows, resulting in defective control and requiring the installation of a positioner.

3.3.16. *Installation of a control valve*

Pneumatic valves are equipped with a positioner that can be purely pneumatic or electro-pneumatic.

Pneumatic positioners amplify the air pressure between the pneumatic control and the valve while comparing the real position and the position required by the control and the real position. The positioner eliminates any discrepancy.

Electro-pneumatic positioners convert the regulator's electric signal into compressed air pressure and, as above, compare the position required and the real position, eliminating any discrepancy.

Positioners compensate either unexpected mechanical friction or a pressure imbalance on the plug of a plug valve.

Compressed air is required for the operation of the actuator and pneumatic regulators. Consumption is in the order of 1.5 $Nm^3.h^{-1}$ of air for each.

Actuator air pressure is between 1.2 and 1.4 barG. If it is equipped with an amplifier, the amplifier supplies the servomotor with a pressure of 1.4–5 bars.

The dew point of air at the highest use pressure should be 5°C below the lowest ambient temperature (or process temperature if this is lower).

This dew temperature is often arbitrarily taken as equal to –40°C. We obtain this result by passing air through a molecular sieve bed (once referred to as silica gel). The installation must include thermal or pneumatic adsorbent regeneration.

If the servomotor is electromagnetic, typically this is simply a front-back motor.

Control valves can be equipped with a bypass. In such cases, we fit block valves upstream and downstream of the valve, together with the bypass. If the servomotor fails, the control valve is isolated and the bypass then allows for manual operation of flowrate control.

Two criteria let us assess whether a bypass is required:

– cost;

– requirement for the process involved.

Thus:

1) If financial resources are very limited, we do not fit a bypass.

2) If financial resources are abundant, we fit all of the control valves with a bypass.

3) A compromise can be as follows :

– Not to fit a bypass on valves for which the ND is below 150 mm.

– To fit on valves whose circulation cannot be interrupted without significant inconvenience (safety or production shutdown).

3.4. The process parameters of a control valve

3.4.1. *Variation in line pressure*

Variation of fluid pressure along a line results from two terms:

1) a dynamic term (always resistant)

In turbulent regime:

$$\Delta P_D = f\frac{\rho V^2}{2}\frac{L}{D} \quad (\text{Pa})$$

ρ and V are, respectively, the density ($kg.m^{-3}$) and the velocity ($m.s^{-1}$) of the fluid. L and D are the length and diameter of the pipe (meters). f is the friction coefficient, which for a pipe is typically in the order of 0.03.

In laminar regime:

$$\Delta P_D = 32\mu V\frac{L}{D^2} \quad (\text{Pa})$$

μ is the liquid viscosity (Pa.s)

During a change in the flow without a change in regime, the new drop in pressure (index 2) is derived from the previous (index 1) as follows:

turbulent regime $\Delta P_2 = \Delta P_1 \times (Q_2/Q_1)^2$

laminar regime $\Delta P_2 = \Delta P_1 \times (Q_2/Q_1)$

2) A static term (driving or resistant)

By agreeing to give a positive sign to this term if it is driving, we write:

$$\Delta P_S = P_s - P_u$$

P_s and P_u are, respectively, the governing pressure in the upstream device (source) and downstream device (use). More precisely:

$$P_s = P_{sG} + \rho_L g H_s \quad \text{and} \quad P_u = P_{uG} + \rho_L g H_u$$

P_{sG} and P_{uG} are the pressures at the liquid surface or simply the pressure of the gas if there is no liquid. H_u and H_s are the liquid levels according to a reference level.

So:

$$\Delta P_S = P_{sG} - P_{uG} + \rho_L g \left(H_s - H_u\right)$$

It follows that we can distinguish two sorts of lines:

– *Natural* driving lines for which ΔP_S is positive and greater than ΔP_D. In these lines, fluid flows by itself in the right direction and the useful pressure ΔP_U is defined by:

$$\Delta P_U = \Delta P_S - \Delta P_D$$

– Naturally, resistant lines in which the fluid does not flow by itself, that is, in which ΔP_S are negative or (if it is positive) its value is less than ΔP_D. In order to make the fluid circulate, the installation of a turbomachine is required (pump, compressor or simply ventilator). In such cases, we have a line that was modified to be driving.

We discount cases in which the device is expected to be volumetric, as the flow obtained would only be slightly dependent on the pressure profile of the line and the intended flow would be obtained by means of the motor rotation speed rather than the control valve.

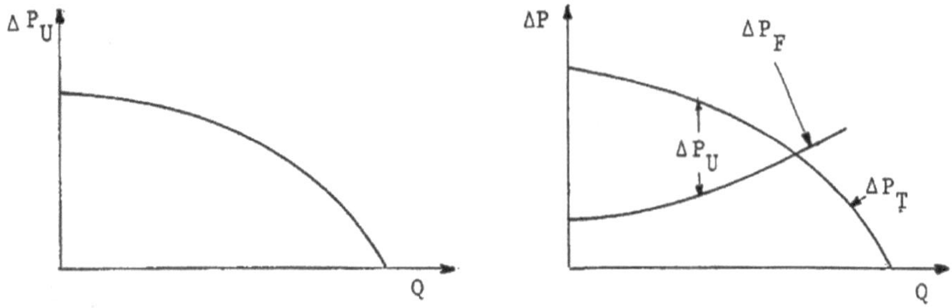

Figure 3.20. *a) Natural driving line; b) made driving line*

For such driving lines, we define:

– the pressure drop for the final flow of fluid:

$$\Delta P_F = \Delta P_D - \Delta P_S$$

– the driving pressure of the turbo machine (pump etc...): ΔP_T

– the usable pressure then becomes:

$$\Delta P_U = \Delta P_T - \Delta P_F = \Delta P_T + \Delta P_S - \Delta P_D$$

Figures 3.20(a) and (b) demonstrate the above elements.

3.4.2. *Disturbances on a line*

A transient variation of ΔP_U can arise. This can be ascribed to two main causes:

– Variation in the pressure of the gaseous phase (whether there is liquid or not) upstream or downstream, which is announced by a global disturbance that is often estimated at 0.05 P_{sG}, where P_{sG} is the upstream gas pressure (source).

– Variation in the fluid level upstream or downstream around normal (operating) level. The maximum amplitude of this variation is:

$$\delta H = \left[\delta\left(H_s - H_u\right)\right] > 0$$

In conclusion, the maximum amplitude for the disturbances is:

$$\delta P = \rho_L g \delta H + 0.05\ P_{sG}$$

3.4.3. *Definition of control valves (principles)*

The basic principle is that the useful pressure must be entirely dissipated by the control valve.

However, valve opening is variable between zero and 100%. If it is closed, the pressure drop is infinite, and if it is wide open:

– in turbulent regime, the pressure drop is proportional to the square of the flow;

– in laminar flow, the pressure drop is simply proportional to the flow.

For the maximum flow Q_M and for the minimum usable pressure, the valve should be wide open and, finally, the applicable principle is as follows:

At maximum flow, the minimum value of useful pressure, and only this value, must be dissipated by the valve in wide-open position.

This parameter avoids oversizing the valve and thereby committing unnecessary investment.

We will examine how to apply this principle to the types of lines that we have defined.

– For a naturally driving line, the pressure drop P_{VO} of the wide-open vent will be:

$$\Delta P_{VO} = \Delta P_U - \delta P$$

– For a driving line driven by a turbomachine, we begin by selecting the value given in the Table 3.1 for ΔP_{VO} according to the valve type:

Valve type	ΔP_{VO} (bar)
Pinch	0.01
Ball	0.05
Butterfly	0.08
Double plug	0.5
Simple plug	0.7
Eccentric shut-off	0.7

Table 3.1. *Pressure drop in valves*

The values above are only applicable for liquids, while for gases they require multiplication according to the following ratio (in international units):

$$\left(\rho V^2\right) \text{gas} / \left(\rho V^2\right) \text{liquid} \# \ \rho_G \times 100/1500 = \rho_G/15$$

Of course, the above is only applicable in turbulent regime and in the expressions above, ρ_G is the gas expressed in kg.m^{-3}.

By knowing ΔP_{VO}, we can obtain the driving pressure of the turbomachine:

$$\Delta P_t = \Delta P_{VO} + \delta P + \Delta P_F \qquad [3.5]$$

NOTE.–

A clogged filter or pipe is announced by an additional irreversible pressure drop ΔP_I and the expressions above become:

$$\Delta P_{VO} = \Delta P_U - \delta P - \Delta P_I$$

$$\Delta P_T = \Delta P_{VO} + \delta P + \Delta P_F + \delta P_I$$

EXAMPLE (study of a line).–

Let us consider the line represented in Figure 3.21.

Pump P_1 transfers a fluid from tank R_1 to tank R_2. Along the line is filter S_1 for which the pressure drop in nominal flow is 0.1 bar, if it is not clogged, and 0.2 bar if it is clogged. The flow through this filter is turbulent. The fluid level in R_1 can vary from 3 to 6 m above the pump axis level. The line proceeds on a pipe bridge whose level is 10m above the pump axis level.

The installation's nominal flow is:

$$Q_N = 25\ m^3.h^{-1}$$

The O.R. required for the installation is such that:

– Oversizing S_u is 120 %

– Flexibility S_o is 40%

The requisite O.R. is accordingly:

$$\frac{S_u - S_o}{S_u} = \frac{120-40}{120} = 66\ \%$$

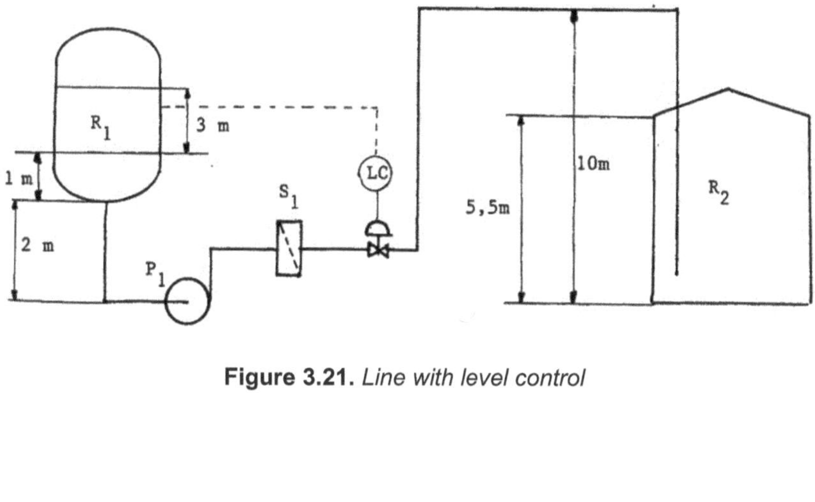

Figure 3.21. *Line with level control*

A simple plug valve is appropriate.

The pressure drop in the pipe for 25 $m^3.h^{-1}$ is 0.3 bar.

The free surface of the liquid in R_1 and R_2 is at atmospheric pressure. The density of the liquid is taken as equal to $kg.m^{-3}$.

The control pump is used to maintain the level in R_1. When the level in R_1 is at its maximum, the pump will have maximum flow.

Maximum flow is:

$$Q_M = 1.2 \times 25 = m^3.h^{-1}$$

For this flow, the dynamic pressure drop in the pipe is:

$$\Delta P_D = (0.3 + 0.2)(30/25)^2 = 0.72 \text{ bar}$$

In both tanks, the distance separating the highest possible level from the lowest possible level is:

– For R1: 3 m;

– For R2: 5 m.

Normal level is 70% of these heights.

The line to treat is the line made active by pump P1. It was not naturally active, as on start-up the siphon effect was not present for the pipe bridge.

Consequently, the liquid should be elevated to 10 m independently of reservoir R2, so that:

The possible drop in $H_u - H_s$ is:

$$0.7 \times 3 = 2.1 \text{ m}$$

The possible increase is:

$$\delta H = 0.3 \times 3 = 0.9 \text{ m}$$

The maximum amplitude of disturbances is:

$$\delta P = 800 \times 9.81 \times 0.9 \times 0.05 \times 1$$

$$\delta P = 0.12 \text{ bar}$$

With the pressure drop across the valve taken as equal to 0.7 bar, its C_v will be at a maximum flow of 30 $m^3.h^{-1}$:

$$C_v = 1.156 \times 30\sqrt{\frac{0.8}{0.7}} = 37$$

Which is a K_v of 370.

The static pressure at the pump's normal suction level is:

$$P_s = (1 + 2 + 0.7 \times 3) \times 800 \times 9.81 \times 10^{-5} = 0.4 \text{ bar}$$

The static discharge pressure is:

$$P_u = 1 + 10 \times 9.81 \times 800 \times 10^{-5} = 1.785 \text{ bar}$$

Hence:

$$\Delta P_S = 0.4 - 1.785 = -1.385 \text{ bar}$$

$$\Delta P_F = 0.72 + 1.385 = 2.1 \text{ bar}$$

When the clogging effect is accounted for in the pressure drop from the filter, the driving pressure of the pump will be, according to equation [3.5]:

$$\Delta P_T = 0.7 + 0.12 + 2.1 = 2.92 \text{ bar}$$

That is a manometric height equal to:

$$2.92.10^5 / (800 \times 9.81) = 37.2 \text{ m.C.L.}$$

NOTE.–

Standard practice dictates that we stack the pressure drops upstream and downstream of the pump. The driving pressure is deducted from the difference.

4

Electric Motors: Performance and Choice of Pumps and Fans

4.1. Choice of motor

4.1.1. *General*

The most widespread motor used in the chemical industry is the asynchronous motor.

4.1.2. *Installed power*

The power rating of motors that we see on the market is standardized up to 132 kW. This power rating is given in Table 4.1:

0.55	3	15	55
0.75	4	18.5	75
1.1	5.5	22	90
1.5	7.5	30	110
1.8	9	37	132
2.2	11	45	

Table 4.1. *Standardized power rating of motors (kW)*

If an engine requires shaft power P_a, we increase this power by 20% and choose the standardized power threshold in the range immediately above. If the required power exceeds 132 kW, the motor will need to be specially conceived by the manufacturer. Motors can be manufactured without difficulty up to a power in the order of 15,000 kW.

The nominal power or installed power is indicated on the identification plate of the motor. Clearly, the motors selected according to the method described here will operate at approximately 3/4 of their nominal load.

4.1.3. *Supply voltage*

Up to 132 kW, the accepted voltage is 380 V. In order to reduce thermic loss in high-power motors, we supply them with high voltage: 3.3 kV, 8.5 kV or 10 kV, according to the characteristics of the electric energy available in the establishment.

The transition between low voltage (380 V) and high voltage applies for power ratings between 132 and 250 kW.

4.1.4. *Rotation velocity*

The rotation velocity of asynchronous motor is, to the nearest 3%:

$$N = \frac{60\ f}{p}$$

N: rotation velocity ($rev.mn^{-1}$)

p: number of pole pairs, typically 1 or 2

f: frequency of alternating current (50 Hz).

Thus, pumps can turn, depending on the circumstances, at 1,500 $rev.mn^{-1}$ or 3,000 $rev.mn^{-1}$, the highest velocity corresponding to the highest rate of wear. With one pair of poles, we have:

$$N = \frac{60 \times 50}{1} = 3000\ rev.mn^{-1}$$

4.2. Utilization of motors

4.2.1. *Starting*

The electric intensity that flows through the motor together with torque that the engine develops is largely dependent on the rotation velocity, as indicated in Figure 4.1.

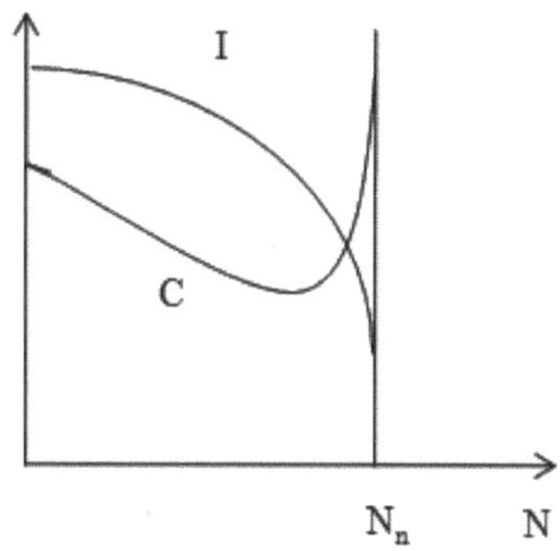

Figure 4.1. *Intensity and torque of an asynchronous motor according to rotation velocity N*

We observe that the ratio of electric intensity on motor starting to that in normal operation is between 4.5 and 8, which on average is equal to 6.

As for torque, it only attains its maximum value when close to the nominal velocity N_n.

Nonetheless, a distinction should be made between low-power motors (380 V) and high-power motors (that are supplied with voltages of several kilovolts).

1) Low-power motors:

– motors are started by simple application of the nominal voltage. The driven engine still has an inertia that is characterized according to its PD^2 (measured in $kg.m^2$).

Starting duration is given by:

$$\tau = \frac{2.75.10^{-6} \times PD^2 \times N^2}{P_n \left[a \frac{C_d}{C_n} - 0.4 \frac{P_a}{P_n} \right]}$$

τ: Starting duration (s)

PD^2: engine moment of inertia ($kg.m^2$)

N: engine rotation frequency ($rev.mn^{-1}$)

P_n: nominal power of motor (kW)

C_d/C_n: ratio of motor starting and nominal torques.

Typically, $C_d/C_n \sim 2$

a: coefficient dependent on the number of pole pairs p

p	a
1	0.8
2	0.85

P_a: engine shaft power (kW)

Thus, in the case of a fan, for example:

$N = 3000\ rev.mn^{-1}$ $\quad PD^2 = 2\ kg.m^2$

$P_a = 1.5\ kW$ $\quad P_n = 2.2\ kW$

$C_d/C_n = 2.3$

$$\tau = \frac{2.75.10^{-6} \times 2 \times 3000^2}{2.2 \left[0.8 \times 2.3 - 0.4 \times \frac{1.5}{2.2} \right]}$$

$$\tau = 14\ s$$

2) High power motors:

high-power motors require an inrush current to start, which can affect the rest of the electric installation. Any attempt to attenuate this effect would be detrimental to motor yield.

Consequently, in order to reduce this current inflow, we must reduce the supplied voltage by adding appropriate equipment on the electric circuit. The device chosen is typically fitted inside the starter box, which could be more expensive than the motor itself.

When a motor is equipped with a switch between one and two pole pairs, it must start at low velocity.

4.2.2. *Power consumption of an installation*

We can choose between two methods to obtain this:

1) The global method (fast and approximate):

assumes that the power consumption is quite simply 80% of the total installed powers P_{Ii}

$$P_{consumption} = 0.8 \times \sum_i P_{Ii}$$

2) The detailed method (slower and more precise):

we make use of the fact that the yield η of an electric motor at 3/4 power can be estimated using the expression:

$$\eta = 0.95 - 0.32 e^{-0.12 P_I} \qquad (P_I \text{ in kW})$$

Knowing the installed powers P_{Ii}, it is easy to deduce the yield of each motor.

If P_a is the shaft power required by the engine and, consequently, supplied by the motor, the electric power consumption P_E is:

$$P_E = P_a / \eta$$

And for the whole installation, power consumption is estimated as follows:

$$P_C = \sum_i \left(\frac{P_{ai}}{\eta_i} \right)$$

4.3. Turbopumps

4.3.1. *The main types of turbopump*

The oil, chemicals and food industries, in hydrodynamic terms, typically use:

– centrifugal pumps, also known as process pumps;

– propeller pumps, also known as axial pumps.

Let us examine these engines.

4.3.2. *Centrifugal pumps*

In number, they represent more than 80% of installed pumps. This is certainly why they are referred to as process pumps.

The suction pipe is horizontal and the discharge is perpendicular to wheel periphery. Various layout norms set the position of the entry and exit flanges for process pumps.

These pumps are simple, easy to maintain, have a small footprint and have a continuous lifespan that can exceed ten years. Discharge height for a single wheel is typically between 10 and 60 m liquid column. Possible flow rates vary between several dozen liters per hour and 1,000 $m^3 \cdot h^{-1}$ with most between 10 and 100 $m^3 \cdot h^{-1}$.

The most primitive type of centrifuge pump comprises a wheel equipped with radial blades; however, it has been observed that in such configurations, turbulence, and consequently energy loss, is considerable. Turbulence intensity, that is the ratio of mean quadratic fluctuations in velocity to the mean of these velocities, has a value close to one. This is the design that is used for radial stirring turbines whose intended purpose is indeed to create the maximum turbulence.

In order to rectify the flow in the wheel of a centrifuge pump, the vanes are arrayed in a spiral. One of them is represented in Figure 4.2.

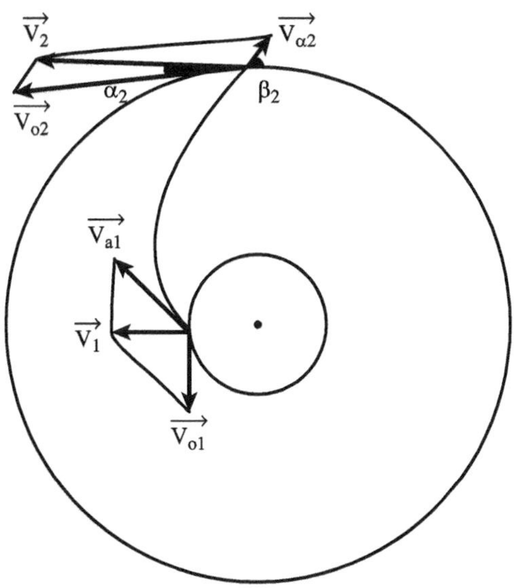

Figure 4.2. *Wheel of a centrifuge pump*

Vector $\overrightarrow{V_2}$, representing velocity relative to the recipient of the liquid exiting the wheel, is the sum of vector $\overrightarrow{V_{o2}}$ orthoradial velocity (periphery) of the wheel and of velocity $\overrightarrow{V_{a2}}$ acquired by the liquid relative to the wheel by following a path between two guide vanes:

$$\overrightarrow{V_2} = \overrightarrow{V_{o2}} + \overrightarrow{V_{a2}}$$

The vane profile is such that on entry of a liquid into the wheel, vector $\overrightarrow{V_1}$, the velocity of the liquid relative to the recipient, runs through the rotation axis. The kinetic moment of the entry fluid is consequently zero relative to this axis, and the creation of kinetic momentum on wheel exit is:

$$(\text{mass flowrate}) \times (\text{orthoradial component of acquired velocity}) \times (\text{lever arm})$$

That is (see Figure 4.2):

$$C = Q\,\varrho V_2 \cos\alpha_2 R$$

Q: volumetric flow rate ($m^3.s^{-1}$)

ϱ: liquid density ($kg.m^{-3}$)

R: wheel radius (m)

This expression is equal to torque C applied to the fluid by the wheel and the power consumption is the result of this torque by the angular rotation velocity.

$$P = C\omega = 2\pi NC = 2\pi NQ\varrho V_2 \cos\alpha_2 R$$

Consequently, the driving pressure communicated by the pump on the liquid is:

$$\Delta P = \frac{P}{Q} = 2\pi N\varrho V_2 \cos\alpha_2 R$$

But, in the exit parallelogram, we observe that:

$$V_{o2} = V_2 \cos\alpha_2 + V_{a2} \cos\beta_2$$

By eliminating $V_2 \cos \alpha_2$ between these last two equations, we get:

$$\Delta P = 2\pi N\varrho R\left(V_{o2} - V_{a2} \cos\beta_2\right)$$

But, the radial exit velocity of the liquid is:

$$V_{a2} \sin\beta_2 = \frac{Q}{2\pi Rl}$$

ℓ: wheel thickness (m)

Furthermore, peripheral wheel velocity is:

$$V_{o2} = 2\pi RN$$

Finally, the driving pressure is:

$$\Delta P = 4 \times \pi^2 \varrho R^2 N^2 - \frac{N\varrho Q}{l \tan\beta_2}$$

Angle β_2 is in the order of 40°.

Practice dictates that we define a pump by flow Q and the manometric height H that it generates:

$$H = \frac{\Delta P}{\varrho g} = \frac{4\pi^2 R^2 N^2}{g} - \frac{NQ}{g l \tan\beta_2}$$

The theoretical power is:

$$P = Q\,\Delta P = 4\pi^2 \varrho\, R^2 N^2 Q - \frac{N\varrho Q^2}{l \tan\beta_2}$$

Note that the expression of H allows us to determine the radius of wheel R corresponding to the given values of H and Q. However, for this, the pump's hydraulic yield η_H must be taken into account.

$$\eta_H = \frac{\text{H practical}}{\text{H theoretical}}$$

Hence:

$$R = \frac{\sqrt{g}}{4\pi N} \sqrt{\frac{H}{\eta_H} + \frac{NQ}{g\ell \tan\beta_2}} \qquad [4.1]$$

This expression can have at least two advantages:

– when we reduce the diameter of a wheel to give it a lower manometric height;

– when we calculate the Reynolds number of the wheel in order to deduce the yield of a pump treating viscous liquid.

The hydraulic yield of centrifugal pumps increases with engine size ($\eta = 0.8$ for the largest pumps) and a value of 0.7 is widely estimated for nominal conditions. However, the internal regulations of a process can cause the flow to vary from 60% to 130% of nominal. A plug valve more or less throttles the process flow in the exit pipe. This is why a yield of 0.6 is preferable.

While the flow/height characteristic is clearly the key element in choosing a pump, the second most important is the net suction head required by the engine to operate without cavitation. To be more concise, we will refer to this height as NPSH.

An engine's available NPSH is the difference between the entry pressure and the vapor pressure of the liquid at that location, all divided by the sum of acceleration due to gravity and the fluid density:

$$\text{NPSH} = \frac{P_1 - \pi_1}{\varrho g}$$

P_1 is the pressure exerted by the fluid on the walls of the suction pipe.

If the liquid is in part pumped from a tank in which it is in equilibrium with its vapor, the available NPSH is obtained by application of the Bernoulli equation:

$$\frac{\pi_o}{\varrho g} + h_o = \frac{P_1}{\varrho g} + \frac{V_1^2}{2g} + h_1 + \frac{\delta P}{\varrho g}$$

Index o refers to the free surface of the tank and δP is the pressure drop due to pipe friction. Accepting that there is no temperature variation between the tank and the pump, then:

$$\pi_o = \pi_1$$

and:

$$\mathrm{NPSH} = \frac{P_1 - \pi_o}{\varrho g} = h_o - h_1 - \left(\frac{V_1^2}{2g} + \frac{\delta P}{\varrho g} \right)$$

This available NPSH must be greater than the NPSH required for operating a pump without cavitation. According to the Euler theorem, the pressure of a fluid in a tube decreases as its velocity increases together with the effect of friction. Accordingly, on wheel entry, there is indeed acceleration and a decrease in vapor pressure that can potentially become negative. Vapor (or dissolved gas) bubbles then form. The duration of bubble formation is in the order of a millisecond, creating a shockwave that is detrimental to the performance of metallic surfaces (see section 4.3.13).

The required NPSH should be provided by the manufacturer and varies according to the square of rotation velocity, which is logical since, in turbulent flow, the pressure drops due to friction are proportional to the square of velocity. Thus, the required NPSH is multiplied by four if we have a pump turn at 3,000 rev.mn^{-1} rather than at 1,500 rev.mn^{-1}.

The required NPSH is subject to a great deal of attention during the installation of new devices, since, in the chemical industry, liquids are frequently conveyed at temperatures close to their boiling point.

If the supplier delivers a centrifugal pump equipped with its motor, the pump and the motor should have been set on a metal chassis beforehand, and the manufacturer ensuring a suitable alignment of the pump axes and motor. When the motor is bought separately, the buyer has to ensure that this alignment is performed correctly

4.3.3. *Centrifugal pump yield*

In a pump, there are two types of energetic losses: hydraulic losses (which correspond to liquid impacts and friction along the pipe and through the pump), and mechanical losses (which are less significant than hydraulic losses).

Pump yield accounts for all these losses, being better when the pump is larger in size, that is, when its flow is higher. Yield may be assessed, for an initial estimation, using the following expression for a liquid whose viscosity is close to that of water:

$$\eta_o = \frac{0.8\,Q + 25}{Q + 55}$$

Q: nominal flow ($m^3.h^{-1}$).

Where the viscosity of the conveyed liquid is high, the yield will be lower, and the value η_o must be multiplied by the coefficient K_v below 1 that evaluates to part of the pump's Reynolds number:

$$Re = \frac{Q\rho}{\mu D}$$

Q: flow ($m^3.s^{-1}$)

ρ: density of liquid ($kg.m^{-3}$)

D: wheel diameter (m)

μ: liquid viscosity (Pa.s (1 Pa.s = 1000 cP))

The table below gives the yield correction coefficient K_v according to the Reynolds number.

Re	K_v
10,000	1
7,000	0.9
5,000	0.8
1,000	0.7
500	0.6
200	0.45
100	0.3

Table 4.2. *Yield correction coefficient*

EXAMPLE 4.1.–

$Q = 31\ m^3 h^{-1} = 0.0086\ m^3 s^{-1}$ $\rho = 800$

$H = 30$ m C.L. $\mu = 200\ cp = 0.2$ Pa.s

$N = 1500\ rev.mn^{-1} = 25\ rev.s^{-1}$ ℓ (estimated) $= 0.04$ m

$tg\ \beta_2 = 0.84\ (\beta_2 = 40°)$

$$\eta_o = \frac{0.8 \times 31 + 25}{31 + 55} = 0.56$$

Wheel diameter is, according to equation [4.1]:

$$R = \frac{\sqrt{9.81}}{4\pi \times 25} \times \sqrt{\frac{30}{0.56} + \frac{25 \times 0.0086}{9.81 \times 0.04 \times 0.84}}$$

$R = 0.074$ m

$D = 2R = 0.158$ m

We note that, in the expression of the wheel radius R, the term under the radical including parameters ℓ and tg β_2 is negligible and, in fact, it is sufficient to write out:

$$R = \frac{\sqrt{g}}{4\pi N}\sqrt{\frac{H}{\eta}}$$

$$Re = \frac{0.0086 \times 800}{0.2 \times 0.158} = 218$$

Hence, by interpolation:

$K_v = 0.46$

And finally, the actual pump yield is:

$\eta = 0.56 \times 0.46 = 0.26$

Strictly speaking, it is this yield value that should have been used to calculate R, resulting in both a higher result and Reynolds number, while K_v would have been lower. Accordingly, our approach is conservative in nature.

4.3.4. *Normal–emergency centrifugal pump systems*

The two pumps, the operating pump and the emergency pump, are set up parallel to one another. Each is isolated by upstream and downstream block valves to allow for intervention on one of the engines, and perhaps even to replace it, even with the pipes full of liquid.

A failure of the operating pump is announced by low pressure on the manometer installed at the common discharge of the pump pair. If the ongoing process cannot be interrupted then the emergency pump must be activated as soon as possible. There are two configurations possible depending on whether the block valves are manual or automatic.

1) Manual block valves:

the upstream block valves are kept open at all times. If the pressure falls, the operator shuts the discharge valve of the pump in operation due to its failure. The operator activates the emergency pump and then progressively opens the discharge valve of this pump. This procedure is known as *closed valve starting*, which avoids motor overcharge due to the abrupt effects of the liquid's mechanical inertia.

The operator can by no means intervene instantaneously, and the discharge pipe (which, by definition, is at the higher pressure) may well back up through the pump, which leads to several drawbacks including:

– siphon draining due to a pipe bridge being crossed, which, in order to be refilled on reinitialization, requires further effort from the emergency pump;

– if the driving pressure is significant, the downstream device will quickly empty into the upstream device and the conveyance of the liquid will need to begin again, hence a loss of time and a disturbance to the process. In particular, the backed up liquid can pollute the upstream liquid.

Such drawbacks can be alleviated by the installation of a no-return valve between each pump and its downstream isolation valve. Nonetheless, such a valve is useless if the system in question is closed-circuit, since, in such cases, the liquid immobilizes itself when pumping stops.

If the pumped liquid is hot (temperature above 70°C), no-return valves are fitted with low-diameter bypasses, so that the emergency pump does not abruptly heat up down on start-up.

If the pumped liquid is charged, the no-return valve will not be sealed but will, however, suffice to significantly slow any reflux of the discharge line.

2) Automatic block valves:

while the upstream valves are manual and permanently open, the downstream valves are automatic. In case of lowering pressure, the discharge valve of the operating pump is closed, and immediately afterwards, the emergency pump is started with the closed valve. The downstream valve of this pump opens progressively over ten seconds.

Owing to the automatic operation of the valves, it occurs quickly and without the need for check valves.

Limit switches (non-mandatory) ensure that the pump line transfer occurs correctly.

3) If pressure has not returned to normal when 10 seconds have gone by since the opening of the downstream valve of the emergency pump, then something has occurred on the line other than a simple operating pump failure. Pumping must then be interrupted so that a thorough verification of the line can be performed.

4.3.5. *Liquid flow criteria*

A pump failure equates to a flow stop, and the only clear criterion of sound operation is the measurement of this flow.

However, as it is less expensive to install a manometer when compared with a flow sensor on pump discharge, there is a tendency to replace the flow measurement with a pressure measurement. We will accept this practice here.

The pump must overcome the resistance pressure, which is the total of a hydrostatic term and dynamic friction term along the line. Consequently, in order for the pressure measurement to announce flow in a significant way, it is necessary that:

$$\frac{\text{friction term}}{\text{friction term} + \text{ static term}} > 0.4$$

Where this is not the case, we must use discharge pressure, the ΔP difference between discharge pressure and suction pressure.

4.3.6. *Safety*

In case of operator failure or automatic system failure, no-return valves should be installed to eliminate the possibility that a pump delivers into another, which is then made to turn backwards (like a turbine). Indeed, the mechanical incidents would be serious on starting up such a pump (shaft twisting and even breaking).

4.3.7. *Drive shaft outlet seal*

Relative to the wheel of a centrifuge pump, the aspirated liquid arrives on the wheel side and the corresponding pipe is in the extension of the wheel axis.

On the other side of the wheel, the shaft is coupled to the electric drive motor. It follows that the wheel shaft must cross the pump body casing, which means that a device is required to ensure the seal between the pumped liquid and the outside environment. This seal can fall into one of the two categories: mechanical seals or compressed braid seals.

With mechanical seals, two disks face each other. The fixed disk moves with the stator while the mobile disk turns with the shaft. Typically, the hardness of the disks differs. For instance, a disk pair could use one graphite disk (soft) and one porcelain disk (hard).

Lubrication is indispensable, both for contact between the shaft and compression braid seals and the disks of a mechanical seal. This means that lubrication liquid must be provided.

The most straightforward solution involves using the system of simple compression seal (that is, one piece), whose lubrication is ensured by the process liquid itself, which proceeds to drain to the outside, contributing to the *fluid loss* in an installation.

However, if the process liquid is toxic, inflammable or particularly fugitive, this solution must be rejected. This is also the case where the process liquid includes suspended particles greater than 10 μm in size. Indeed, such particles would deteriorate the seal by scratching the surface or creating preferential pathways between the shaft and the seal. In such cases, an outside liquid should be used, which is often utility water.

If the process liquid can tolerate the addition of utility water, then a simple seal is sufficient. The lubrication liquid is introduced from the outside, flowing to the inside of the pump. If, on the other hand, mixing is to be avoided, a double-seal system needs to be employed. The double seal is essentially a barrage between the process liquid and the lubrication liquid. This lubrication liquid can come from an external reservoir fitted with a level alarm or simply connected to the utility water network.

4.3.8. *Cooling requirement*

In general, the temperature of the bearings oil must not exceed 80°C, hence, when appropriate, the need for cooling in the bearing chamber.

If the temperature of the process liquid exceeds 180°C, the pump base must also be cooled to avoid alignment defects between the pump and motor.

Finally, if the temperature of the process liquid is above 200°C, the mechanical seals must also be cooled.

4.3.9. *Dry running centrifugal pumps*

This situation may arise if the upstream capacity empties unexpectedly. In order for a centrifugal pump to run dry for a short time without significant inconvenience, the following three conditions must be met:

– the bearings must be greased independently from the process fluid (presence of an external oil reservoir);

– the seals must be lubricated by another liquid than the process liquid. The required flow is from 100 to 200 $L.h^{-1}$ and the pressure of the lubrication fluid (often water) must, in normal operation, be greater than the discharge pressure. In order to allow for unexpected events, we set this at 1.2 times discharge pressure. Compression braid seals must be avoided here as if they are not lubricated uniformly, they can heat and be damaged;

– vibration tests on the empty pump must be performed by the manufacturer prior to engine purchase.

A pump capable of dry running does not require a low-level alarm on the upstream capacity.

4.3.10. *Hermetically sealed centrifugal pump*

Traditional centrifugal pumps include a braid (or mechanical seal) that stops leaking liquid at the shaft passage through the pump casing (this passage for the shaft is located on the opposing side to the suction inlet). This shaft outlet allows for the machine to be driven by a motor.

When dealing with liquid ammoniac, or toxic liquids (sodium cyanide solution) the seal, which is inevitably imperfect, is unacceptable. This is why we have recourse to hermetically sealed centrifugal pumps. These pumps come in two kinds:

– wet rotor pumps;

– magnetic drive pumps.

In both systems, the motor stator is immovably with the pump casing.

Drowned rotor pumps are covered with an impervious envelope that turns with it, through which the magnetic field created by the stator acts. Liquid moves around the stator and the rotor, thereby cooling both.

In magnetic drive pumps, liquid circulates around the inside of the rotor envelope and only cools the rotor (whose envelope is fixed). The stator can be cooled by cooling fins. These pumps are less expensive than drowned rotor pumps.

4.3.11. *Propeller pumps*

These pumps not only are capable of high flows (up to 10,000 $m^3 \cdot h^{-1}$) but also generate modest discharge heights (6–10 m C.L.). They are used in circulation loops. The pump case is designed to form an elbow. Their yield can be estimated at 0.7.

4.3.12. *Turbopump shaft power*

This is given by the following expression:

$$P_a = \frac{Q \times \rho \times g \times H}{3600 \times \eta \times 1000}$$

P_a: shaft power (kW)

Q: flow conveyed ($m^3.h^{-1}$)

ρ: fluid density ($kg.m^{-3}$)

g: acceleration due to gravity (9.81 $m.s^{-2}$)

H: manometric height that the pump must be capable of (m)

η: yield.

EXAMPLE 4.2.–

$$Q = 31\ m^3/h \qquad H = 30\ m\ C.L.$$

$$\rho = 800\ kg/m^3 \qquad \eta = 0.26$$

$$P_a = \frac{31 \times 800 \times 9.81 \times 30}{3600 \times 0.26 \times 1000}$$

$$P_a = 7.8\ kW$$

4.3.13. *Cavitation [MUT 01]*

Let us consider the apparition of vapor bubbles within a liquid.

The chemical potential of the liquid and of the vapor within a bubble is:

$$\mu_v = \mu_v(P_o, T_o) + kTLn\left(P_L + \frac{2\gamma}{r}\right) - kTLnP_0$$

$$\mu_L = \mu_L(P_o, T_o) + v_L(P_L - P_o)$$

v_L: molecular volume of liquid (m^3.molecule^{-1}).

Going from a liquid state to a vapor state, the free enthalpy spent for the formation of a whole bubble containing n molecules is:

$$\Delta G_B = \int_o^n \left[\mu_V(n) - \mu_L\right] dn$$

Let us integrate these parts, recalling that μ_L does not depend on n:

$$\Delta G_B = \left[\mu_V(n) - \mu_L\right] n - \int_o^n n d\mu_V(n)$$

Now, we must assess the integrand. Pressure within the bubble is:

$$P_V = P_L + \frac{2\gamma}{r}$$

γ: surface tension of the liquid (N.m^{-1})

r: bubble radius (m).

As the bubble grows, P_L remains constant and:

$$dP_V = -\frac{2\gamma}{r^2} dr$$

The chemical potential inside the bubble varies by $d\mu_V$. This is the potential of n molecules within the bubble.

$$n = \frac{4}{3}\frac{\pi r^3}{v_V} = \frac{4}{3}\frac{\pi r^3}{k_B T} P_V$$

k_B: Boltzmann constant (J.molecule^{-1})

v_V: molecular volume of vapor (m^3, molecule^{-1})

The chemical potential of the vapor molecules is:

$$\mu_V = \mu_V^o + kTnP_V$$

Therefore:

$$d\mu_V = kT\frac{dP_V}{P_V} = \frac{-2\gamma}{r^2}\frac{kT}{P_V}dr$$

Hence:

$$nd\mu_V = -\frac{8}{3}\pi\gamma r dr \text{ and } \int_o^n nd\mu_V = \frac{-4\pi\gamma r^2}{3}$$

Finally:

$$\Delta G_B = \frac{4\pi\gamma r^2}{3} + k_B Tn\left[\mathrm{Ln}\left(\frac{P_L + \frac{2\gamma}{r}}{P_o}\right) + \frac{v_L(P_o - P_L)}{k_B T}\right]$$

But, accounting for the fact that:

$$n = \frac{4\pi r^3 P_V}{3k_B T} \quad \text{and} \quad \frac{dn}{dr} = \frac{4\pi r^2 P_V}{k_B T} - \frac{4\pi r^3}{3k_B T} \times \frac{2\gamma}{r^2}$$

we find:

$$\frac{d\Delta G_B}{dr} = 4\pi r \left[\text{Ln}\left(\frac{P_L + \frac{2\gamma}{r}}{P_o} \right) + \frac{v_L (P_o - P_L)}{k_B T} \right] \left(P_L r + \frac{4}{3}\gamma \right)$$

Let us find the values of r for which $\frac{d\Delta G}{dr} = 0$

– $r = 0$ is a trivial solution;

– the cancellation of the square bracket corresponds to the thermodynamic balance;

– the cancellation of the bracket corresponds to the critical radius r_c:

$$r_c = -\frac{4\gamma}{3P_L}$$

This solution is only possible if the *liquid pressure P_L is negative.*

When r is greater than the value r_c, $\frac{d\Delta G_B}{dr}$ becomes negative and, spontaneously, r increases. We need the contribution of newly vaporized molecules. This is the problem of *cavitation*, whose explosive character is well documented. Cavitation can tear metallic particles in contact with the liquid from the wall, which can *severely damage* a pump wheel, boat propeller or hydraulic turbine wheel. A study on erosion by cavitation in centrifugal pumps may be found in the work of Gülich and Rösch [GÜL 89].

4.4. Volumetric pumps

4.4.1. *Need for volumetric pumps*

Centrifugal pumps should notably be replaced with volumetric pumps in two significant instances:

1) the product is viscous (oil) or even positively thick (food products such as mayonnaise). Indeed, the yield of a small centrifugal pump can decrease

to 0.2 if the viscosity of the liquid reaches 200 cP; 80% of the motor energy will then be lost heating the product, which can be unacceptable;

2) a high discharge pressure is required (500 bars).

The installation of volumetric pumps demands certain precautions:

– provide a discharge safety valve in case of a unanticipated problem in the liquid flow;

– the seals must be sufficient. Indeed, as the discharge pressure increases, the internal return fluid flow increases. When the fluid flow reaches 40% of the theoretical flow, we can say that the pump is poorly used.

Owing to the pressure drop along a pipe increasing with the product's viscosity, the two preceding criteria often occur simultaneously.

4.4.2. *Piston pumps*

Piston pumps are recommended for high pressures (500 bars) and non-abrasive suspensions. Otherwise, the size of solid particles should be below 10 μm. In such cases, we use a pump with a driving piston.

Pump flowrate is connected to the piston displacement velocity. In triangle O A B in Figure 4.3, we describe the classic relationship:

$$L^2 = x^2 + R^2 + 2x\,R\cos\alpha$$

This second-degree equation in x admits the solution:

$$x = -R\cos\alpha + \sqrt{R^2\cos^2\alpha + L^2 - R^2}$$

If A is the piston cross-section, the theoretical flow is:

$$Q = A\frac{dx}{d\tau} = A\omega\left[R\sin\alpha + \frac{R^2\sin\alpha\cos\alpha}{\sqrt{R^2\cos^2\alpha + L^2 - R^2}}\right]$$

With: $\omega = d\alpha/d\tau$.

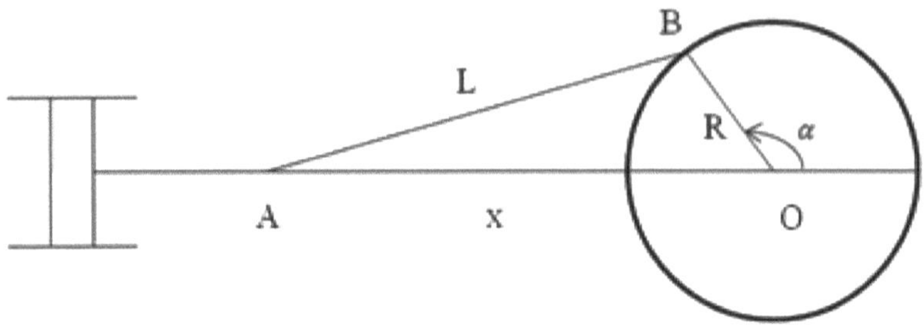

Figure 4.3. *Piston displacement*

Flow is not constant, depending on α. Figures 4.4 and 4.5 express Q according to α for:

– a simple-effect piston;

– a double-effect piston.

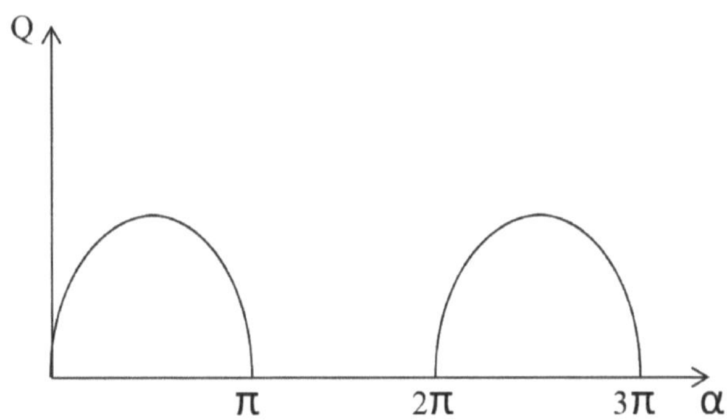

Figure 4.4. *Simple-effect piston*

Note that the flow of a double-effect piston is less irregular than that of a simple-effect piston. We can further improve a continuous flow by using a number of cylinders greater than one and shifting their operating angle α. However, engines such as this are expensive and complex. Pump discharge can be managed in a more straightforward manner with a single effect cylinder pump and a tank under constant pressurized gas since, on leaving this tank, flow will be regular.

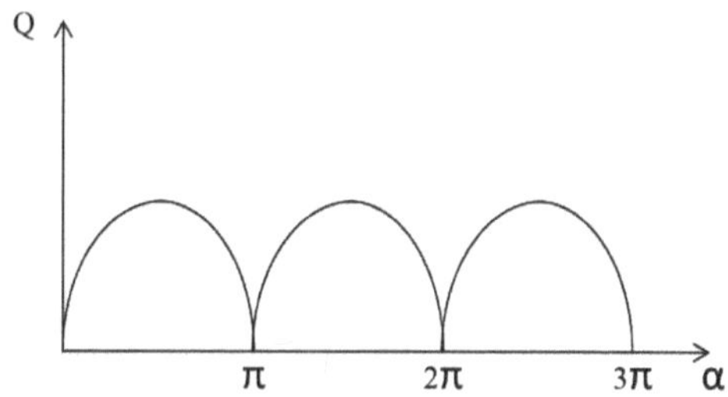

Figure 4.5. *Double-effect piston*

The average flow of piston pumps can reach 150 $m^3.h^{-1}$, but large engines are rare. However, all dosing pumps are piston pumps whose weak flow (sometimes only several $cm^3.h^{-1}$ s) can be controlled by action on the piston stroke.

The mechanical resistance of certain piston pumps allows them to withstand variable discharge pressures according to the engines, of several dozen bars to thousands of bars.

With equal flow, the yield of a piston pump is comparable to that of a centrifugal pump, although piston pumps are more expensive.

4.4.3. *Gear pumps*

These pumps are used for viscous liquids like oils and can convey 70 $m^3.h^{-1}$ of liquid at a pressure in the order of 150–180 bars.

Flow fluctuations due to the gears may lead to vibrations.

4.4.4. *Moyno pumps*

The rotor, resembling a sort of twisted sausage in stainless steel, turns in an elastomer stator, which is cylindrical and twisted in a similar way but hollow.

This type of volumetric pump practically avoids any pulsation. It is suitable for very viscous products or non-abrasive sludge. It is widely employed in the food industry, namely for vegetable, animal, mineral oils, fat, glucose and sucrose solutions, molasses, paper pulp, paint, varnish, mayonnaise, jam, soap, ink, etc.

At 200 rev.mn^{-1}, some of these pumps can discharge 200 m^3.h^{-1} under 5 bars and, at 1,800 rev.mn^{-1}, another type of Moyno pump can discharge 3 m^3.h^{-1} under 45 bars.

4.4.5. *Double-screw pump*

The Moyno pump stator cannot withstand some products such as hot bitumen at 140°C. In such cases, we use a double-screw pump that works somewhat like an extruder and can create pressures of 170 bars.

This pump has the advantage of being capable of operating in both directions, which can be useful to fill and empty a tank.

4.4.6. *Sealed volumetric pumps (membrane)*

These engines do not have sealing devices and are well suited to fugitive, toxic and pollutant liquids.

If the membrane is actioned by a metal stem, the pressure created can reach 70 bars. The flow is below 200 m^3.h^{-1}.

When a suspended solid must not be exposed to high shearing effects (flocculation effluents), we use membrane pumps derived by compressed air. In such cases, discharge pressure does not exceed 5 bar (air compressors typically discharge at 7 bar).

Membrane pumps have the remarkable property of being able to dry run without requiring any special precautions. Indeed, other engines introduce the mutual friction of solid surfaces, which necessarily requires lubrication.

4.4.7. *Volumetric pumps and net suction pressure*

Just as we defined the NPSH for centrifugal pumps, we must establish the net NPSH required for a volumetric pump. Indeed, on passing through the suction valves, and more generally, on entry into any pump, there is a drop in pressure that can result in cavitation. Consequently, a membrane pump discharging 20 bar must be supplied with a pressure of 2 bar abs., which requires the upstream installation of a centrifuge-type feeder pump.

A Moyno pump could also be equipped with a feeder screw for its suction. In which case, the NPSH is not only connected to the vapor pressure of the product, but also to its viscosity (or its consistency).

4.4.8. *Flow regulation of volumetric pumps*

While this regulation is performed by throttling the fluid process on turbopump discharge (axial and centrifuge), volumetric pump flow regulation is performed by varying the rotation velocity of the pump by means of an electronic or purely mechanical device, with the least expensive system being the variator belt used for powers below 300 kW. For higher power levels, a more elaborate system is required.

4.4.9. *Shaft power*

We can estimate yield at 0.7.

Power is then:

$$P_a = \frac{Q \times \Delta P \times 10^5}{3600 \times \eta \times 1000} = \frac{Q\Delta P}{36\eta}$$

Q: flow conveyed ($m^3.h^{-1}$)

ΔP: pressure increase due to pump (bars)

η: yield

EXAMPLE 4.3.–

$$\Delta P = 60 \text{ bar} \qquad Q = 10 \text{ m}^3.\text{h}^{-1}$$

$$P_a = \frac{10 \times 60}{36 \times 0.7} = 23.8 \text{ kW}$$

4.5. Special cases

4.5.1. *Pumps for liquid–gas mixes*

These are pumps capable of conveying a dispersion of gas in a liquid. A traditional centrifugal pump is not suitable for this as, under the effect of centrifugal force, the liquid gathers in the volute and the rotor becomes isolated in a gas pocket.

Only volumetric pumps are self-priming. However, there are a variety of pumps having a starwise rotor that have the advantage of producing a continuous flow, accepting gas–liquid mixes and being relatively inexpensive. The paddles on the wheel are flat and radial; suction and discharge are vertical, situated on the upper part of the pump and spaced around one wheel diameter. But this pump does not work like a paddle pump. Essentially, the volume available for the conveyance of emulsion is a lateral toric channel with an approximately semi-circular cross-section that is formed in the stator of one side of the wheel, and which follows the wheel periphery. Emulsion viscosity, while relatively low, is sufficient for wheel movement to draw the emulsion present in the lateral channel and make it run, from the entry to the exit, a significant part of rotor circumference.

Strictly speaking, these pumps are neither centrifugal nor volumetric. Rather, they are viscous driving. They can be used to pump the waste waters from a sump or drainage tank, as the suction pipe, located below the pump level, is often empty of liquid. However, there needs to be liquid in the pump for starting up. These pumps, whose NPSH is no less than classic centrifugal pumps, are the specialty of the SIHI.

4.5.2. *Self-priming pumps*

To draw liquid from a sump or a ditch, we can use the pumps known as submersible rotors. This refers to the fact that the rotor is permanently submerged in the liquid. The axis of these pumps is vertical and very long, with the motor at the ground level.

We can also use totally submersible pumps (motor included), which are brought to the surface for maintenance by means of a chain. These pumps are less expensive than the former. Of course, both the pump and the motor are encased in a hermetically sealed casing.

No preparation is required for starting-up these pumps.

4.5.3. *Slurry pumping*

Slurry is any solid suspension greater than 50 μm in size and comprising less than 17% of solids in volume. Such slurries are frequent in crystallization operations. We establish this content limit for two reasons:

– the energy used for the transportation of slurry must be reasonable;

– the thermic transport coefficient decreases significantly if the solid limit exceeds 20%.

A suspension containing more than 20% of solid in volume is known as sludge. Sludge is frequent in ore processing stemming from the underflow of gravity decanters (and also centrifuges).

Slurries containing less than 7% solids in volume are treated with half-open wheels, that is, with only one flange. For levels between 7 and 17%, the wheel must be totally open, and if the solid is abrasive, further precautions must be taken:

– peripheral wheel speed must remain below 1,300 $m.s^{-1}$ in order to minimize volute wear;

– sufficient space must be kept clear between the wheel and the case;

– an abrasion-resistant material may be required (15% silicon cast iron or rubberlined steel).

4.5.4. *Sludge pumping*

Sludge is a solid dispersion such that the apparent viscosity measured on a gradient of 5 s^{-1} is in the order of a kilopascal.second (kPa.s). Sludge can be divided into two distinct categories:

1) with a concentration of more than 17% by volume, if the solid particles are larger than 50 μm in size;

2) with a concentration only above 1 or 2% by volume, if the size of the solid is less than 10 μm. This often applies to biological suspensions including bacteria (which are rod-shaped and approximately 10 μm in length) or yeast (which are spherical in form and approximately 1 μm in size). These suspensions readily form sludge, and we must bear in mind that biological organisms are composed of 85–95% water and only 1% dry matter can correspond to 20% of solid in terms of volume. Talc and activated carbon, and numerous pigments, also have an average size below 10 μm and readily become sludge due to their high volumetric area, *absorbing and fixing a large proportion of the water present*, so that the fraction of effective liquid volume is less than an elementary calculation would have us believe.

Sludge is typically non-Newtonian. Its flow in a pipe is accompanied by a drop in pressure that can reach several dozen bars. Consequently, the choice of volumetric pump should be made according to its mechanical resistance.

NOTE.–

The distinction between slurry and sludge obtained from powders (particles less than 50 μm in size) is best categorized according to criteria of consistency rather than of apparent viscosity. However, we will not address this here.

4.5.5. *The water screw*

It can be useful for a water treatment plant to bring water up from one basin to another, the level of which may be around two meters higher.

In such cases we use a screw, whose axis makes an angle of approximately 30 sexagesimal degrees to the horizontal.

The flow of this screw is given by the expression:

$$Q = \mu N D^3$$

D: exterior diameter of propeller (m)

d: shaft diameter

$$d/D = 0.5$$

p: propeller pitch:

$$p/D = 1$$

L: propeller length

$$L \# 1.5D \text{ or } 2D$$

N: rotation velocity ($rev.mn^{-1}$)

$$N = \frac{50}{D^{2/3}}$$

Q: flow ($m^3.h^{-1}$)

μ: flow coefficient

$$\mu = 4.7$$

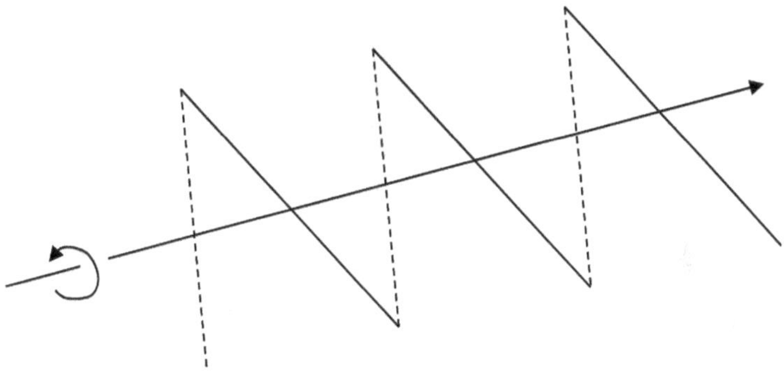

Figure 4.6. *Water screw*

EXAMPLE 4.4.–

$$D = 1 \text{ m} \qquad p = 1 \text{ m}$$

$$N = \frac{50}{1} = 50 \text{ rev.m.n}^{-1}$$

$$Q = 4.7 \times 50 \times 1 = 235 \text{ m}^3.\text{h}^{-1}$$

The volumetric yield η can be defined by:

$$Q = \eta \times 60 \times \pi D^2 \times p \times N \times 0.5$$

Since the water level in the screw is horizontal, channel filling is divided by two.

In the previous example:

$$\eta = \frac{235}{60 \times \pi \times 50 \times 0.5} = 0.05 = 5\%$$

This is a mediocre yield as the water tends to descend along the channel under the influence of gravity.

4.6. Fans

4.6.1. *Similarity laws and electrical power consumed by a fan*

Where:

Q: flow in real volume ($m^3.s^{-1}$)

D: propeller diameter (m)

P_a: theoretical aerodynamic power (Watt)

H: manometric height attained (m G.C, meters *gas column*)

N: rotation velocity ($rev.mn^{-1}$)

$$\frac{Q_2}{Q_1} = \frac{N_2}{N_1} \times \left(\frac{D_2}{D_1}\right)^3$$

$$\frac{H_2}{H_1} = \left(\frac{N_2}{N_1}\right)^2 \times \left(\frac{D_2}{D_1}\right)^2$$

$$P_a = H\rho_G g \times Q$$

$$\frac{\rho_{a1}}{\rho_{a2}} = \left(\frac{N_2}{N_1}\right)^3 \times \left(\frac{D_2}{D_1}\right)^5$$

ρ_G: density of gas exiting fan

g: acceleration of gravity (9.81 $m.s^{-2}$)

The fan calculation is performed using a velocity triangle as for pumps (see Figure 4.2).

Fans can be centrifuges (like the pumps of the same name) or rather axial propellers (more often).

Electrical power consumed is:

$$P_c = \frac{P_a}{1000\ \eta_v \eta_M}$$

η_M: motor yield: never less than 0.85

η_v: aerodynamic yield of ventilator: more than 0.7

P_c: electrical power consumed (kW)

4.6.2. *Fan flow calibration*

1) By variation in velocity using:

– a motor of variable speed;

– a mechanical or hydraulic variator (if a wide range of flow is required);

– transmission by variable diameter pulley.

2) By modification of the aeraulics system;

– a discharge register reduces flow by creating an additional drop in pressure;

– the titled blades, placed opposite to the aspiration inlet, fixed on radial axes and driven to rotate by a shared command. The inlet blades provide the fluid with a pre-rotation movement around the fan axis and in the direction of the fan, whose operation is thereby modified. Consequently, the driving height and flow vary according to the inlet blades configuration.

3) Advantages of the inlet blades:

– easy installation;

– low cost;

– good energy yield;

– good stability.

4.6.3. *Uses of fans outside of pneumatic transport*

It is worth noting that humans require 1 kg of air per hour. The maximum acceptable values of air in $g.m^{-3}$ are:

CO	Cl_2	HCl	SO_2	CO_2	SiO_2	Dust
0.01	0.003	0.02	0.03	0.12	0.0001	0.001

The energy required to heat industrial premises with a 6 m high ceiling is 14 kJ per m^3, per degree Celsius of temperature difference between the inside and outside temperatures and per hour.

A coal household's consumption of combustible air is in the order of 15 m^3 per kg of coal burned.

Comfort temperatures	°C
Houses, restaurants, theaters, cinemas	20
Workshops	18
Hospitals	22
Silos (potatoes, apples)	4
Grain silos	18
Cheese aging facility	4
Wine cellar	15

Table 4.3. *Comfort temperatures*

4.6.4. *Ventilator start time*

The time required for a ventilator to start is given by the following expression, which is used by constructers:

$$T_{dem} = \frac{5.1.10^{-6} PD^2 N^2}{2aP_n - \dfrac{0.4P_a}{\eta_v}} \quad \text{(in seconds)}$$

PD^2: ventilator moment of inertia ($kg.m^2$)

N: rotation velocity (1,500 or 3,000 $rev.mn^{-1}$)

P_n: nominal power of motor (kW)

P_a: theoretical aerodynamic power of ventilator (kW)

η_v: aerodynamic yield; above 0.65

Coefficient a characterizes the asynchronous motor

Number of poles	Coefficient a
2	0.8
4	0.85
6	0.9
8	1

Table 4.4. *Time coefficient*

EXAMPLE 4.5.–

$N = 2900\ rev.mn^{-1}$ $\quad PD^2 = 1.9\ kg.m^2$

$P_n = 1.5\ kW$ $\quad P_a = 1.2\ kW$

$A = 0.8$ $\quad \eta_v = 0.7$

$$T_{dem} = \frac{5.1.10^{-6} \times 1.9 \times 2900^2}{2 \times 0.8 \times 1.5 - \dfrac{0.4 \times 1.2}{0.7}}$$

$$T_{dem} = 47.5 \text{ s}$$

4.6.5. *Noise caused by a ventilator*

The sound power created is in the form:

$$W = kQP^2$$

W: Watt

Q: flow ($m^3.s^{-1}$)

P: aerodynamic pressure created by the ventilator (mm C.E).

The level of sound power is measured in decibels (dB)

$$N_W = K + 10\ \log_{10} Q + 20 \log_{10}\left(\frac{P}{100}\right)$$

The constant K, supplied by the manufacturer, decreases as the size of the device increases. Its value varies between 35 and 71.

Let us recall that:

$$1 \text{ atm} \# 10000 \text{ mm CE} \# 10^5 \text{ Pa}$$

Hence:

$$1 \text{ Pa} = 0.1 \text{ mm CE}$$

The level of acoustic pressure is defined by:

$$N_p = N_W - 10 \log_{10} S$$

S is the surface of the sphere, the radius of which is the distance between the source (the engine) and the measurement point. This is measured in m^2.

This expression may be corrected according to:

– the type of fan;

– the configuration of the environment.

We will not detail how to perform these corrections.

EXAMPLE 4.6.–

$Q = 13\ m^3.s^{-1}$ $\quad P = 50$ mm CE $\quad K = 40$ dB

$$N_W = 40 + 10\log_{10} 13 + 20\log_{10} \frac{50}{100}$$

$$N_W = 40 + 11.14 - 6.02$$

$$N_W = 45\ \text{dB}$$

5

Polymer Extruder Screw

5.1. Introduction of extrusion screw

5.1.1. *Extruder principle*

An extrusion screw is a screw that turns in a barrel transporting a product in the form of solid particles toward a nozzle; little by little the particles melt along the screw. Polymer extrusion screws are enclosed in a cylindrical barrel rather than in a trough (launder) like bulk solid transporting screws. For satisfactory operation, the product must slide along the screw flights, which consequently must be polished. Moreover, the friction coefficient along the barrel must:

– be weak in the longitudinal direction (axial);

– be high in the perpendicular direction.

This is why the barrel must be equipped with grooves in the axial direction, which avoids the product turning in the same direction as the screw.

The barrel can be double walled, which allows for both water cooling or heating with steam along the barrel.

The screw axis comprises three successive work zones that will be addressed here.

Subject to requirements, the ratio of screw length to diameter varies between 15 and 30.

5.1.2. *Geometrical description of a screw*

Consider a full cylinder with circular cross-section. We will call this the screw shaft.

Around this shaft, we roll a slim and deformable ribbon in a spiral so that at each point of contact between the ribbon and the shaft, the tangent plane of the ribbon, must be perpendicular to the tangential plane of the shaft. This ribbon is the screw *flight*, the two sides of which are the flight *flanks*.

The helical line that ensures contact between the slim ribbons of the shaft is called the flight root. The spiral line located at a distance from the axis equal to the exterior radius of the flight is the outer helix.

When the flight has performed one turn, it has axially progressed by a length equal to the screw *pitch*.

If we cut the helix by a plane containing its axis, we get a series of grooves along the axis which are section of that which is known as the *channel*.

The channel *bottom* is the shaft. The *edges* of the channel are the flight *flanks* offset by one full rotation. The channel *depth* is measured from the *exterior diameter* of the helix and lengthwise, that is along the screw axis.

The screw turns in a hollow cylinder that we call the *barrel*. The radial distance separating the surface of the barrel and the exterior perimeter of the crew is called the *clearance*.

The slim and deformable ribbon that we have used in the description above in reality is not slim and its thickness can be in the order of a tenth of the channel width, at least in the case of an extruder screw due to reasons of mechanical resistance.

The polymer fills the channel. It forms a *cord* of rectangular cross-section. On entry, the bed is made of solid particles. After melting of a fraction of these particles, the bed is split into two contiguous beds (one solid and the other liquid) that continue at the same velocity in the channel. We will see that the liquid cord is located on the pushing side of the channel.

Let us name ϕ the angle that the outer helix forms with the reference plans (perpendicular to the screw axis).

Unwind the helix from a tangential plane parallel to the axis and on the length of one turn, that is πD, where D is the exterior diameter of the screw. We can also unwind the flight root by one turn on the length πd, where d is the shaft diameter. In both cases, we have progressed axially by a distance equal to the screw pitch.

For the pitch of the helix, we can write:

$$p = \pi D \operatorname{tg} \phi_D = \pi \, d \operatorname{tg} \phi_d$$

Any plane perpendicular to the axis will for present purposes be known as the reference plane. Let us note that for a square pitch p = D, $\operatorname{tg} \phi = 1/\pi$ and $\phi = 17.65°$.

We observe that the screw angle ϕ *increases as we approach the axis* reaching $\frac{\pi}{2}$, if the shaft diameter d nears zero. The generally accepted value corresponds to the outer helix.

Eccher and Valentinotti [ECC 58] measured the velocity of a viscous liquid in the channel and in the three directions in space. Figure 7 of their work shows a view in space of the velocity for the three following cases:

– maximum flow;

– intermediate flow;

– zero flow.

These results confirm the approach calculations of Carley *et al.* [CAR 53].

In addition, these velocity measurements show that, along the channel, the liquid is driven by a spiral movement that was confirmed in the calculation of Griffith [GRI 62].

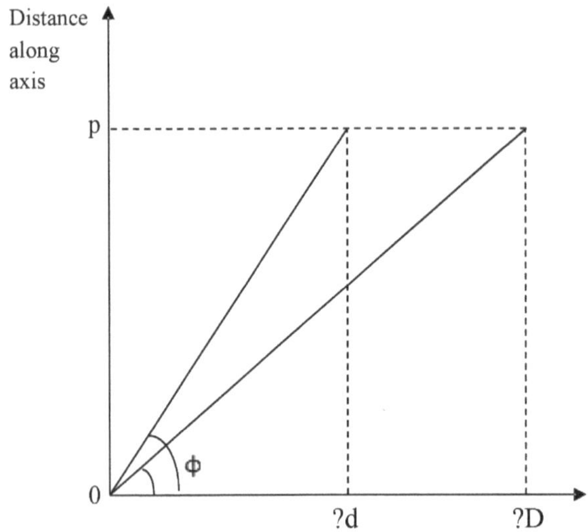

Figure 5.1. *Unwinding the helix through one turn*

Essentially, we can discern three zones spread along the screw:

– feeding in solid particles. The channel depth is constant and must be a little deeper than required to allow for space between the solid particles;

– compression, fusion. The channel narrows in that its depth decreases linearly with the product's progression. This increases the contact surface with the barrel, favoring fusion;

– conveyance of liquid to the nozzle. The channel depth is constant. The mix homogenizes before crossing the nozzle at which pressure is in the order of 200 bars.

The ratio L/D of screw length to diameter is often equal to 24 as there are 12 inches in a foot. However, a ratio of 30 rather than 24 can reduce unmelted. Nonetheless, shaft torsion resistance limits the screw length L.

The clearance between the barrel and the flight is in the order of 25–50 μm, but the thickness of the liquid film separating the barrel from the solids not yet melted can vary from 250 to 600 μm, which corresponds to a velocity gradient of 1,000–3,000 s^{-1}.

The exterior diameter of the screw is, according to requirements, between 2.5 and 8 inches, that is between 6.35 and 20.32 cm.

Channel depth is limited by the mechanical resistance of the flight. Supply depth is in the order of three times the depth on exiting the fusion zone.

5.1.3. *Screw rotation direction*

Take the example of a wood screw. To screw it into wood, it must be turned clockwise. We say that the fight is right threaded.

When the wood screw is applied, we consider the wood along the screw to move away from the screw point.

When we turn a right-threaded screw counter-clockwise, the screw withdraws from the hole it was in. This is exactly the same direction that we have to turn the right-threaded extruder screw so that the product is channeled along this screw toward the die.

If the extruder screw is left-threaded, it has to turn clockwise (to the right).

The flight of a screw has two flanks: a downstream flank and an upstream flank. The nature of these flanks is *constant* along the screw length.

The sides of the channel in which the product is moved are:

– the downstream flank of the flight that is the upstream and driving flank with respect to the channel;

– the upstream flank of the flight that is the downstream and drawing flank with respect to the channel.

The above applies to the sides of the channel for *two successive turns* of the flight.

5.1.4. *Movement of fluid particles*

We will allow for three hypotheses:

1) The barrel is smooth, without friction, and the screw is rough ($\mu = 0.25$).

In this case, the product turns with the screw and does not move along the channel.

2) The barrel presents non-zero friction (for instance, $\mu = 0.35$) and the screw is perfectly smooth ($\mu = 0$).

The rotation movement of the screw leads to a friction force on the barrel F_T that can be split up into:

– an orthogonal force F_N on the flight (this force does not work);

– a parallel force F_τ to the flight that has the product move along the channel.

In real terms, we should add that the dynamic friction of the polymer on the barrel acts not with the force F_T but with a force F_R proportional to the velocity V_R that we present in section 5.2.4.

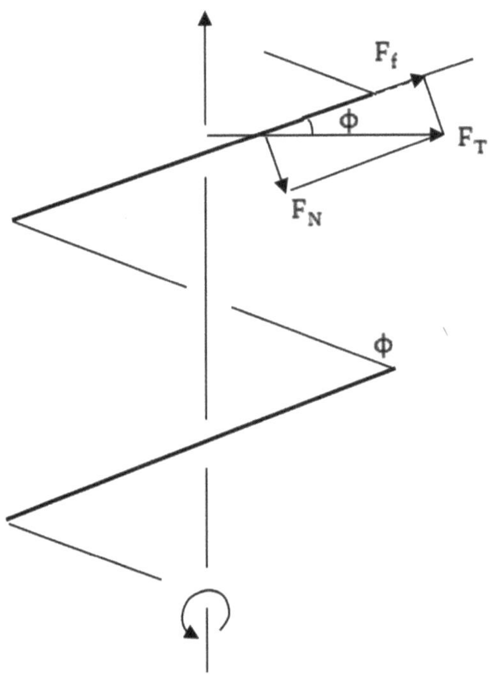

Figure 5.2. *Right-threaded screw*

3) The barrel and the screw present resistance to polymer or divided solid movement.

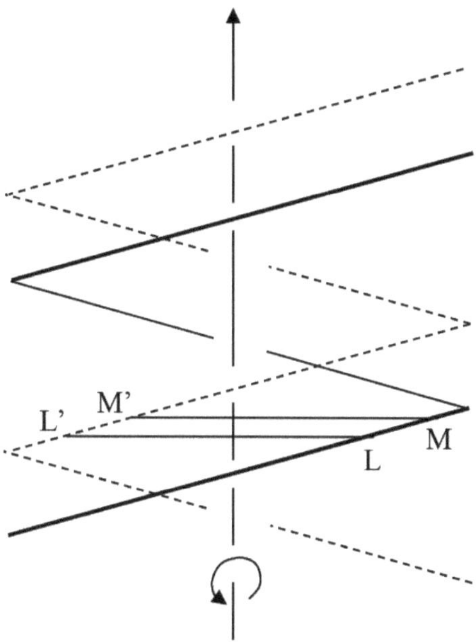

Figure 5.3. *Polymer or divided solids movement*

Following one rotation of the screw by angle α, the flight has moved from LM to L'M', while the product has moved from L to M' having moved along the channel for distance L'M' = LM.

We need to find a relationship between α and LM or rather between L'L = M'M and L'M' = LM. The accepted way of proceeding consists of introducing angle θ that separates the right LM' and the right L'L and M'M. This equates to calling θ the angle of product (divided solids or polymer) with the reference planes (perpendicular to the screw axis).

By writing out the balance of forces and moments relative to the rotation axis, Darnell and Mol [DAR 56], together with Chung [CHU 70], provide a means of determining angle θ. The former uses the dry friction coefficients (with the polymer still in a solid state) and the latter, the theory of viscous friction for melted polymers.

5.1.5. *Preliminary studies*

Carley *et al.* [CAR 53] were the first to propose a flow expression including a term for drag and a term for pressure.

Squires [SQU 58] expresses the slowing of flow by the flight flanks using the F_T and F_P coefficients on each of the two terms of Carley *et al.* [CAR 53]. He also studied the influence of channel curve as well as that of a partially filled channel. He accounted for a transverse force field on the channel and the ratio between channel depth and width.

The velocity measurements of Eccher and Valentinotti [ECC 58] support the calculations of Carley *et al.* [CAR 53].

Mohr *et al.* [MOH 61], together with Squires [SQU 58], account for flight thickness in their flow expression (equation 5). These authors show that the fact a liquid is non-Newtonian does not significantly modify the velocity field.

Booy [BOO 63] studies the influence of channel curve on flow, pressure and required mechanical pressure.

But it was above all Street [STR 61] who showed that the movement of liquid is a result not only of screw rotation but also of the back flank driving on the polymer bed. This drive is initially exerted on the solid particles, compressing them, causing them to melt due in large part to the heat coming from the barrel and transmitted by the driving flank. As the fusion of these particles continues, the air between the particles moves upstream, escaping from the machine.

Normally, due to the cross-section being reduced by the presence of empty spaces, these particles flow faster than the liquid produced by fusion. Melted polymers move more slowly in the channel

NOTE.–

Sometimes agglomerates of softened particles are formed, which can adhere to the metallic walls, provoking a reduction in flow velocity and giving it an irregular aspect. This is one reason to avoid heating the screw and its flight.

5.2. Movement of the polymer in the screw channel

5.2.1. *Introduction*

We will successively examine:

– velocity and flowrate in the channel according to velocity V_c in the upper part of the cord;

– friction on the four sides of the cord;

– the velocity V_c calculation of the cord on the channel surface;

– the effect of the orthoaxial component (perpendicular to channel axis) on dissipated energy;

– leak flow between the screw and the barrel;

– friction on the barrel from flight edges;

– screw shaft power;

– a simple and well-known relationship for extrapolating shaft power when the screw goes from diameter d to diameter D.

The bed of polymer that fills the channel is made of two continuous cords, one liquid and the other solids which move through the channel at the same velocity.

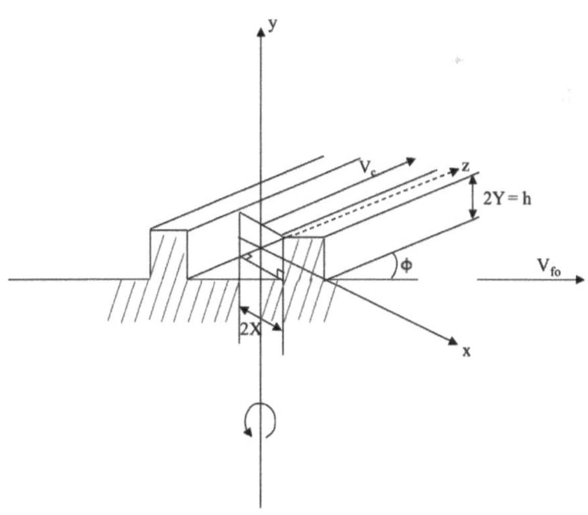

Figure 5.4. *Channel dimensions*

In Figure 5.4, the Oy axis merges with the screw axis.

5.2.2. *The viscosity of extruded polymers*

The energy consumed to bring a volume unit of product from the extruder can be defined as:

$$W_s = \tau \dot{\gamma}^2 \mu$$

τ: residence time of product (s)

$\dot{\gamma}$: shear rate (s^{-1})

μ: apparent viscosity of product (Pa.s)

W_s: energy density ($J.m^{-3}$).

Apparent viscosity μ can be measured with a Severs viscometer comprised of a vertical cylinder partially filled with product and fitted with a pipe in the middle of its inferior face. By applying a known uniform pressure P on the upper part of the product, we can determine the flow coming through the pipe and thereby write out the Poiseuille law:

$$Q = \frac{\pi \, \Delta P r^4}{8 \mu L} \text{ and, for a pseudoplastic } \frac{\Delta P}{L} = 2C\left(\frac{3n+1}{\pi n}\right)^n \frac{Q^n}{r^{3n+1}}$$

L: tube length (m)

P: pressure (Pa)

r: tube radius

n: rheotypical exponent

μ: apparent viscosity (Pa.s)

C: consistency of the pseudoplastic (Pa.s^n (effectively $\tau = C\ \dot{\gamma}^n$))

Q: flowrate (m^3.s^{-1}).

The order of magnitude of apparent viscosity for an extruded liquid (that is, a paste) is:

1,000–2,000 P

or 100,000–200,000 cP

or 100–200 Pa.s

5.2.3. *Movement equation between two parallel plates: velocity and flowrate*

The notations are as follows:

– Oz: axis according to the direction of movement.

– y: distance relative to median plane (2Y is the distance between the plates).

– Ox: axis perpendicular to y and Oz (2X: width of plates).

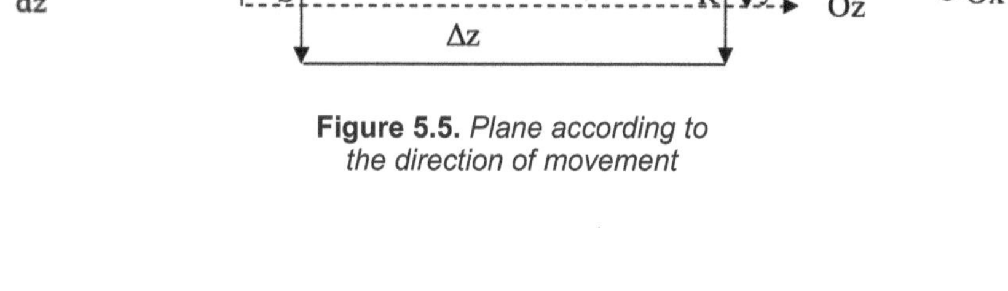

Figure 5.5. *Plane according to the direction of movement*

Let us consider plate PQRS parallel to two plates and length Δz. The sheer stress on its lower part is:

$$C\varepsilon_V\left(\frac{dV}{dy}\right)^n \quad \text{with} \quad \varepsilon_V = \frac{dV/dy}{|dV/dy|}$$

On its upper part:

$$C\varepsilon_V\left[\left|\frac{dV}{dy}\right|^n + d\left|\frac{dV}{dy}\right|^n\right]$$

The resulting force due to sheer is:

$$2X \times dz \times \varepsilon_V d\left|\frac{dV}{dy}\right|^n$$

The force of pressure in the direction -Oz exerted on the plate is:

$$-\Delta P dy \times 2X$$

Equalizing these two forces, we obtain the movement equation:

$$\varepsilon_V d\left[\left|\frac{dV}{dy}\right|^n\right] = \frac{-\Delta P}{C\Delta z}dy \qquad \varepsilon_V\left|\frac{dV}{dy}\right|^n = \frac{-\Delta P}{C\Delta z}y$$

Indeed, the velocity is maximum on the median plane: $\frac{dV}{dy} = 0$ if $y = 0$:

1) The two plates are immobile:

The movement equation is written as:

$$\varepsilon_V\left|\frac{dV}{dy}\right| = \frac{dV}{dy} \text{ and } dV = -\varepsilon_V\left|\frac{\Delta P y}{C\Delta z}\right|^{\frac{1}{n}} dy$$

$$\text{assuming } \varepsilon_p = \frac{\Delta P/\Delta z}{|\Delta P/\Delta z|} \quad \text{and } y > 0$$

$$V(y) = -\varepsilon_P \left|\frac{\Delta P}{C\Delta z}\right|^{\frac{1}{n}} \int_y^h y^{\frac{1}{n}} dy = -\varepsilon_P \left|\frac{\Delta P}{C\Delta z}\right|^{\frac{1}{n}} \left(h^{\frac{1}{n}+1} - y^{1+\frac{1}{n}}\right)\frac{n}{n+1} < 0$$

Indeed, $V = 0$ if $y = h$.

In Figure 5.6, $\frac{\Delta P}{\Delta z}$ is directed toward the right and $V(y)$ toward the left.

Velocity is zero along the linings and maximum in absolute value on the median plane ($y = 0$).

Note that if $\frac{\Delta P}{\Delta z}$ is positive, $\frac{dV}{dy}$ is negative.

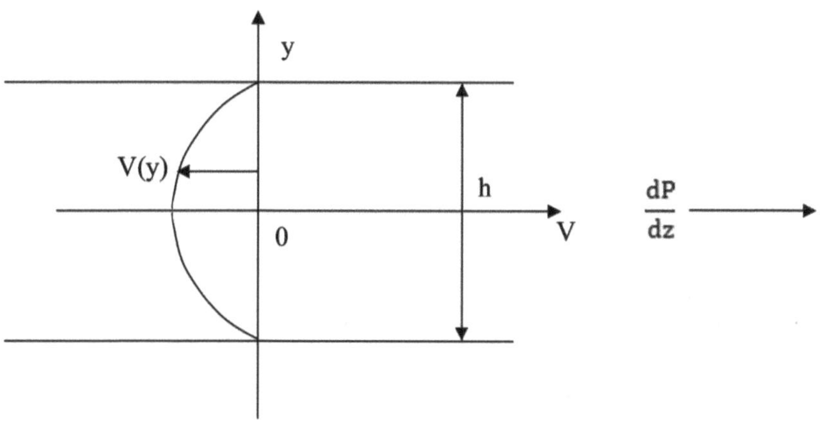

Figure 5.6. *Polymer velocity (two immobile plates)*

The corresponding flow is, if ε_P is taken for $\Delta P/\Delta z$.

$$-\varepsilon_V \left|\frac{dV}{dy}\right| = -\varepsilon_V \left|y\frac{\Delta P}{C\Delta z} + K_1\right|^{\frac{1}{n}} \quad \text{where} \quad \frac{-dV}{dy} = -\varepsilon_V \left|\frac{y\Delta P}{C\Delta z} + K_1\right|^{\frac{1}{n}}$$

$$Q = \frac{-\varepsilon_p 4\, nX}{2n+1}\left|\frac{\Delta P}{C\Delta z}\right|^{\frac{1}{n}} h^{\frac{1}{n}+2}$$

Q, negative, is directed to the left if ε_p is positive.

2) The upper plate moves with the velocity V_C relative to the lower plate:

$$-\varepsilon_V \frac{d}{dy}\left[\left|\frac{dV}{dy}\right|^n\right] = \frac{\Delta P}{C\Delta z}$$

$$-\varepsilon_V \left|\frac{dV}{dy}\right|^n = y\frac{\Delta P}{C\Delta z} + K_1 \text{ of the same sign as } -\varepsilon_V$$

$$-\varepsilon_V \left|\frac{dV}{dy}\right| = -\varepsilon_V \left|y\frac{\Delta P}{C\Delta z} + K_1\right|^{\frac{1}{n}} \quad \text{where} \quad \frac{-dV}{dy} = -\varepsilon_V \left|\frac{y\Delta P}{C\Delta z} + K_1\right|^{\frac{1}{n}}$$

$$V = V_C \quad \text{for} \quad y = \frac{h}{2}$$

$$V = 0 \quad \text{for} \quad y = \frac{-h}{2}$$

$$V(y) = \varepsilon_V \int_{-\frac{h}{2}}^{y} \left|y\frac{\Delta P}{C\Delta z} + K_1\right|^{\frac{1}{n}} dy$$

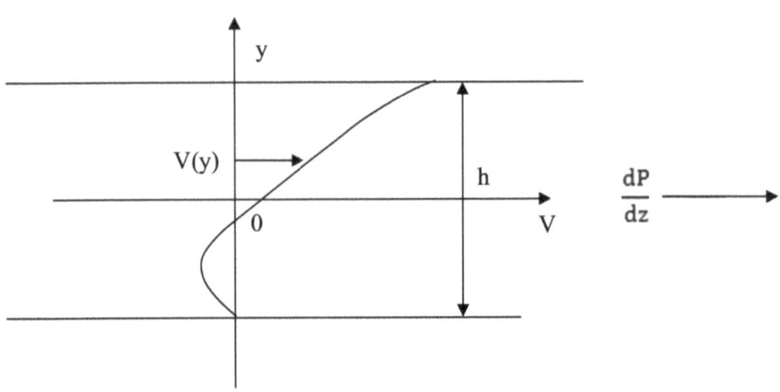

Figure 5.7. *Polymer velocity (a mobile plate)*

A primitive is:

$$\varepsilon_V \frac{C\Delta z}{\Delta P}\left| y\frac{\Delta P}{C\Delta z} + K_1 \right|^{\frac{1}{n}+1} \times \frac{n}{n+1} + K_2$$

$$y = \frac{h}{2} \quad \varepsilon_V \frac{C\Delta z}{\Delta P}\left| \frac{h\Delta P}{2C\Delta z} + K_1 \right|^{\frac{1}{n}+1} \times \frac{n}{n+1} + K_2 = V_c$$

$$y = -\frac{h}{2} \quad \varepsilon_V \frac{C\Delta z}{\Delta P}\left| K_1 - \frac{h\Delta P}{2C\Delta z} \right|^{\frac{1}{n}+1} \frac{n}{n+1} + K_2 = 0 \;\; \Delta K_2 = -\frac{CL}{\Delta P}\left| K_1 - \frac{h\Delta P}{2CL} \right|^{\frac{1}{n}+1}$$

Eliminating K_2:

$$\frac{\varepsilon_V nC\Delta z}{(n+1)\Delta P}\left[\left| \frac{h\Delta P}{2C\Delta z} + K_1 \right|^{\frac{1}{n}+1} - \left| K_1 - \frac{h\Delta P}{2C\Delta z} \right|^{\frac{1}{n}+1} \right] = V_C$$

Let K_1^* be the root of this equation. This root is positive.

Finally:

$$V(y) = \frac{\varepsilon nC\Delta z}{(n+1)\Delta P}\left[\left| y\frac{\Delta P}{2C\Delta z} + K_1^* \right|^{\frac{1}{n}+1} - \left| K_1^* - \frac{h\Delta P}{2C\Delta z} \right|^{\frac{1}{n}+1} \right]$$

Now, let us calculate the flowrate, writing out:

$$I(y) = \int V(y)\,dy = \frac{\varepsilon_V nC\Delta z}{(n+1)\Delta P}\left[\frac{1}{\frac{1}{n}+2}\left| \frac{y\Delta P}{2C\Delta z} + K_1^* \right|^{\frac{1}{n}+2} - y\left| K_1^* - \frac{h\Delta P}{2C\Delta z} \right|^{\frac{1}{n}+1} \right]$$

Let us take this primitive between $\frac{-h}{2}$ and $\frac{h}{2}$:

$$I\left(\frac{h}{2}\right)-I\left(\frac{-h}{2}\right)=\frac{\varepsilon nC\Delta z}{(n+1)\Delta P}\frac{n}{(1+2n)}\left[\left|\frac{h\Delta P}{2C\Delta z}\right|+K_1^*\right|^{\frac{1}{n}+2}-\left|K_1^2-\left|\frac{h\Delta P}{2C\Delta z}\right|\right]^{\frac{1}{n}+2}$$

$$-\frac{\varepsilon_V nC\Delta zh}{(n+1)\Delta P}\left|K_1^*-\frac{h\Delta P}{2C\Delta z}\right|^{\frac{1}{n}+1}$$

This expression is equal to $\frac{Q}{2X}$, where Q is the flowrate required in the channel and X the half-width of the channel.

NOTE.–

Carley *et al.* [CAR 53] give a simplified expression for velocity and flow by accepting that the polymer is Newtonian.

$$V=\frac{V_c y}{h}+\frac{(y^2-hy)}{2\mu}\left(\frac{\Delta P}{\Delta z}\right)$$

$$Q=\frac{\pi^2 D^2 Nh\sin\varphi\cos\varphi}{2}-\frac{\pi Dh^3\sin\varphi}{12\mu}\left(\frac{\Delta P}{\Delta z}\right)$$

These expressions could be useful for starting iterative calculations.

5.2.4. *Polymer friction on a solid lining*

The polymer bed can be in solid or melted form.

If the polymer is solid, along with Chung [CHU 70], we accept that the cord is still coated with a film of melted polymer as it is in contact with a hot wall.

Now we suppose that in both cases the shearing effect on the lining is of the form:

$$\tau = C\,V^n$$

C: coefficient different from consistency

n: rheotypical exponent.

Typically, for melted polymers:

$$0.3 < n < 0.4$$

Note that for plastisols (dispersion of polymer particles in a plasticizer), the exponent n is superior to 1 (plastisols are dilatants) but plastisols are not used in extrusion operations.

Coefficient C can take three values, at least as far as the present subject is concerned.

C_x for the channel sides, C_y for the bottom and C_f for the channel surface.

Value C_f corresponds to the interposition of a film of melted polymer between the wall and the solid polymer. The thickness of this film is in the order of millimeters and, through this thickness, the velocity gradient is in the order of 1,000 s^{-1}.

NOTE.–

Of course, for the melted polymer on the cord sides, we should write:

$$\tau = C\,V_f^n \quad \text{with} \quad V_f^n = \frac{1}{2h}\int_{\frac{-h}{2}}^{\frac{+h}{2}} |V(y)|^n dy$$

Upon contact with the channel bottom:

$$\tau = -\varepsilon_V C\left|\frac{dV}{dy}\right|^n_{y=\frac{-h}{2}} = -\varepsilon_V C\left|K_1^* - \frac{h\Delta P}{2C\Delta z}\right|^n$$

5.2.5. *Cord movement and simulation of screw operation*

We suppose the screw flight is to the right. Consequently, the screw turns to the left. Relative to the screw, the barrel seems to turn to the right (clockwise). Its peripheral velocity is equal to $\pi DN = V_{f0}$. Velocity $\vec{V}_c$ is the polymer velocity on the channel surface.

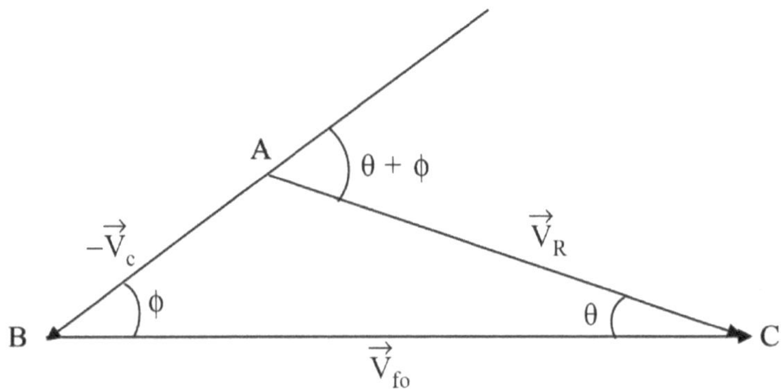

Figure 5.8. *Velocity* $\vec{V}_{fo}$ *of barrel and of bed* $\vec{V}_c$

Relative velocity V_R of the barrel to the cord.

$\vec{V}_R$

Relative velocity $\vec{V}_R$ of the barrel to the cord is:

$$\vec{V}_R = \vec{V}_{fo} - \vec{V}_C = \vec{V}_{fo} + \left(-\vec{V}_c\right)$$

The vector $\vec{V}_c$ is angle ϕ with the planes perpendicular to the screw axis. The (positive) modulus of these vectors are V_R, V_{fo} and V_c, respectively.

In triangle ABC, we observe the following relations:

$$\frac{V_{fo}}{\sin(\theta+\phi)} = \frac{V_c}{\sin\theta} = \frac{V_R}{\sin\phi}$$

Hence:

$$V_c = \frac{V_{fo}\sin\theta}{\sin(\theta+\phi)} \qquad [5.1]$$

$$V_R = \frac{V_{fo}\sin\phi}{\sin(\theta+\phi)}$$

Let us consider a segment of the cord whose dimensions are:

– length: dz;

– width: 2X;

– height: 2Y = h.

On the channel surface, force τ_R exerts on surface dz in direction $\vec{V}_R$:

$$\tau_R = C_R V_R^n \qquad dF_R\cos(\theta+\phi) = 2XdzC_R V_R^n\cos(\theta+\phi)(\text{along the channel})$$

This expression corresponds to viscous friction, where n is the rheotypical exponent of the polymer.

Similarly, the viscous friction along the channel walls (the bottom and the two sides) is expressed by force F_C.

$$\tau_{fr} = C_{fr} V_f^n \quad \text{so} \quad dF_{fr} = (2\,X\,C_X + 4YC_Y)V_c^n dz$$

F_R is a driving force for the cord and F_f is a braking force (of thermic dissipation). We hypothesize that the transversal forces on the channel applied on the cord by both sides are in balance, as both sides are in contact with the liquid or rather with a liquid film interposed between the wall and the solid particles.

The difference between the driving force F_R cos (θ + ϕ) and the friction resistance F_{fr} amount to the bed pressure:

$$dF_R\cos(\theta+\phi) - dF_{fr} = 4XYdP$$

Integrate and divide by Z:

$$2XC_R V_R^n\cos(\theta+\phi) - 2(XC_X + 2YC_Y)V_c^n = 4XY\frac{dP}{dz} = \frac{dF_P}{dz} \qquad [5.2]$$

Thus, we have obtained a relationship between the angle θ (unknown) and the pressure gradient along the channel.

This process must be completed by the calculations of Chung [CHU 70] that incorporate the force typically exerted by the pushing flank on the cord along the channel, which is expressed by a pressure P_n and a force (R−r) dz $P_n = dF_n$.

The author projects the forces exerted (on part of the cord) on the tangential plane to the screw and parallel to its axis. He also expresses the projection of these forces on a reference plane and takes the torque of these projections relative to the screw axis. Between these two equations, he eliminates P_n. This provides an equation which, once added to equation [5.2], allows for the calculation of angle θ and gradient dP/dz. One of these equations including P_n then lets us calculate P_n.

Let us now calculate the screw shaft power. The channel length is equal to L/sin ϕ. The torque of force Fn relative to the axis is:

$$M_{Fn} = \left(\frac{R+r}{2}\right)(R-r)P_n\left(\frac{L}{\sin\phi}\right)\cos\left(\frac{\pi}{2}-\phi\right)$$

The power of mechanical origin of the shaft is:

$$\Pi_m = \frac{\pi DN\left(R^2-r^2\right)P_n L\cos\left(\frac{\pi}{2}-\phi\right)}{2\sin\phi}$$

In fact, the pressure gradient is necessary so that, on exiting the screw, pressure P_{fi} is sufficient for nozzle entry. Thus, the melted polymer can pass through the die. We then have:

$$\frac{dP}{dz} = \frac{P_{fi} - P_0}{Z}$$

P_0: pressure at the feed hopper of the screw feeding shaft (Pa).

Pressure P_0 is equal to 1 bar abs, that is, to atmospheric pressure.

In fact, die pressure depends on the total flow Q_T crossing it. Similarly, gradient $\frac{dP}{dz}$ also depends on the total flow Q_T along the channel (see equation [5.2]).

Therefore, Q_T must satisfy:

$$Q_T = f_{fi}(Q_{fi}) = Q_V \quad \text{with} \quad Q_T = Q_V - Q_P \qquad [5.3]$$

Equation [5.3] allows the calculation of Q_T. We see that the screw simulation is not straightforward, requiring the calculation of three unknowns: θ, dP/dz and Q_T. This is achieved by iterations on Q_T.

NOTE.–

Extruder screws typically comprise three sequential segments that differ in channel depth. The problem may be resolved by writing that at the plane separating two segments pressure and flow are common to these segments.

NOTE.–

There is an orthoaxial component of force F_R that is orthogonal to the channel axis. This component tends to make the bed turn on its axis. The module of this component is $F_R \sin(\theta + \phi)$. Eccher and Valentinotti [ECC 58] provided the experimental confirmation. We will examine this more closely.

5.2.6. *Effect on the cord of the orthoaxial component of relative velocity V_R*

– The bed is entirely in a liquid state. The component of $\vec{V}_R$ is orthogonal to the channel axis (orthoaxial component) and has the effect of giving the liquid a movement of rotation around the channel axis.

The braking pressure on the channel wall is of the form:

$$\tau = C\left|\vec{V}_{fo}\right|^n$$

These stresses form the angle $\frac{\pi}{2}$ with the screw axis.

– The cord is entirely in a solid state. It cannot therefore turn under the influence of the orthoaxial component of $\vec{V}_R$. We will call this component V_{R0} that is entirely consumed in friction on the cord surface. The pressure due to $\vec{V}_R$ is broken down into two parts:

$$\text{axial:} \quad CV_R^n \cos(\theta + \phi) = \tau_a$$

$$\text{orthoaxial:} \quad CV_R^n \sin(\theta + \phi) = \tau_0$$

The orthoaxial stress is immediately consumed as heat. All that remains is the axial stress arising in equation [5.2].

The orthoaxial component of V_R behaves differently in both cases, which perhaps explains why the solid bed breaks as it undergoes fusion.

5.2.7. *Leak flowrate between the flight and the barrel*

This flow diminishes from the calculated flow.

On both sides of the flight, the pressure difference is:

$$\Delta P_{fu} = \frac{(P - P_0)}{Z} \times \frac{\pi D}{\cos\phi}$$

The cross-section available to leak flow is:

$$A_{fu} = L_{he} \times S = \frac{LS}{\cos\phi}$$

The velocity of the barrel relative to the screw is V_{fo}.

Expression [5.3] suffices to calculate the flow.

5.2.8. *Friction of the barrel on the flight edge*

If L is the screw length, the length of the helix on the outer edge of the flight is:

$$L_{he} = \frac{L}{\cos\phi}$$

δ: clearance available between the barrel and the helix

L: screw length.

The velocity gradient on thickness δ is:

$$\frac{V_{fo}}{\delta}$$

V_{fo}: orthoradial velocity of barrel.

The corresponding shear stress is:

$$\tau = C\left(\frac{V_{fo}}{\delta}\right)^n$$

This stress applies to the surface:

$$eL_{hel} = \frac{Le}{\cos\phi}$$

e: flight thickness.

The braking torque due to friction is:

$$M_{fr} = C\left(\frac{V_{fo}}{\delta}\right)^n \frac{Le}{\cos\delta} R$$

R: exterior screw radius.

If $2\pi N$ is the angular rotation velocity of the screw, power Π spent to counter friction along the outer helix is:

$$\Pi_{he} = 2\pi NRC_R \left(\frac{2\pi NR}{\delta}\right)^n \frac{Le}{\cos\phi} = \frac{(2\pi NR)^{1+n} LeC_R}{\delta^n \cos\phi}$$

5.3. Heat for polymer melting

5.3.1. *Melting of polymer particles*

Maddock [MAD 59] and Street [STR 61] studied the manner in which particles of polymer melt. Let us consider a right-threaded screw turning counter-clockwise.

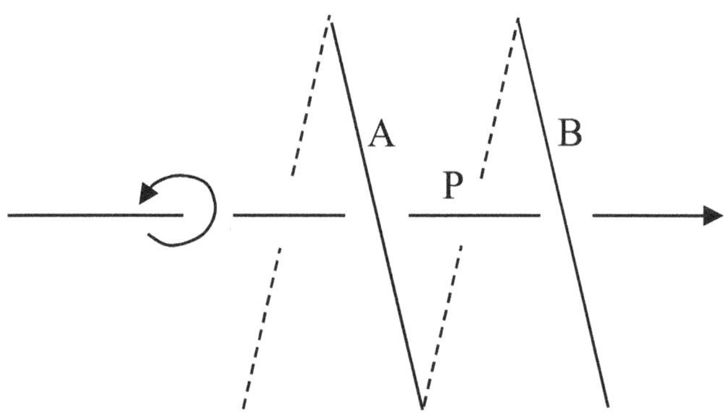

Figure 5.9. *Progression of the polymer*

Let us consider product P located between parts A and B of the flight.

Part B of the thread is the *forward* part relative to the product.

Part A is the part *driving* the product from *behind.*

In Figure 5.10, the solid particles S, still not melted, constitute the cross-hatched section between parts A and B of the flight. This solid part S is not homogenous and is heated on approaching the melted part F.

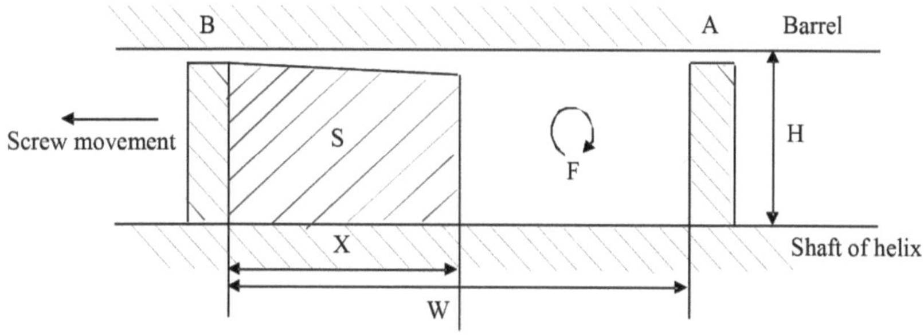

Figure 5.10. *Segment of screw channel*

The melted and circulating part F of the product is in contact with the driving flight A. The barrel slows the progression of the product and causes the melted part to circulate.

Finally, fusion occurs by two mechanisms:

– contact with the hot barrel;

– contact with already melted product;

– heat dissipation between the barrel and the solid.

Above the melted film solid S circulates from B toward A due to the braking effect of the barrel.

5.3.2. *Heat dissipated by viscous friction between two parallel plates in relative movement*

The heat production in m^3 and by second is:

$$q(y) = -\tau_{yz}\left(\frac{dV}{dy}\right)$$

The sheering stress is, if ε_v is the sign of $\frac{dV}{dy}$:

$$\tau_{yz} = -C\left|\frac{dV}{dy}\right|^n \varepsilon_V \qquad \left[\tau_{yz}\right] = \text{Pa}$$

Hence, the volumetric thermal power produced is:

$$q(y) = +C\left|\frac{dV}{dy}\right| n\varepsilon_V \frac{dV}{dy} = C\left|\frac{dV}{dy}\right|^{n+1} \qquad > 0$$

Effectively:

$$[q] = [Pa]s^{-1} = \frac{\text{force.m}}{m^3 \quad s} = \frac{\text{watt}}{m^3}$$

q is always positive. But the volumetric thermal power produced is ΔQ in the volume W L Δy.

$$q(y) = \frac{\Delta Q}{WL\Delta y}$$

2W: band width

L: band length

Δy: band thickness.

All along the height h, the heat produced is:

$$Q(h) = WL\int_0^h q(y)\,dy = WLC\int_{-h/2}^{h/2}\left|\frac{dV}{dy}\right|^{n+1} dy$$

5.3.3. *Thermal dissipation due to the liquid cord rotating by itself*

We accept a *greatly simplified* law for the velocity gradient.

$V_R \sin(\theta + \phi)$ is the superficial orthoaxial velocity of the cord.

The bed radius is taken, on average, as equal to R_m.

$$\frac{X+Y}{R_m} = \frac{X}{Y} + \frac{Y}{X} \quad \text{d'où} \quad \frac{1}{R_m} = \frac{X^2+Y^2}{XY(X+Y)}$$

The velocity gradient is:

$$\frac{dV}{dy} = \text{grad}V = \frac{V_R \sin(\theta + \phi)}{R_m}$$

The energy dissipation for one turn is:

$$R_m \Pi \, D \, C \left| \frac{dV}{dy} \right|^{n+1} p$$

If p is the exterior pitch of the screw, the number of turns will be:

$$n_{tour} = \frac{L}{p} \text{ and for a square screw } \frac{L}{\pi D}$$

Finally, the screw's energy dissipation by rotation of the liquid bed is:

$$n_{tour} \times R_m \pi DC \left| \frac{dV}{dy} \right|^{n+1} p$$

Therefore:

$$Q = LR_m \pi DC \frac{V_R \sin(\theta + \phi)}{R_m}^{n+1} \qquad (\text{Watt})$$

5.3.4. *Preliminary waiting time before fusion as described by Tadmor et al.*

Prior to the initial formation of the liquid bed, the liquid film between the solid and the barrel must have reached a thickness in the order of five times the existing clearance between the barrel and the outside edge of the flight (clearance can be 10 µm for a screw of 20 cm in diameter). The liquid film thickens faster in proportion to barrel heat and the reduction in screw velocity. A thick liquid film is required for the flight to *sweep* the melted polymer along the barrel surface and direct it to the liquid cord.

After a certain waiting time, we can accept that the volume of liquid formed is equal to that of the empty voids in the first particle layer. The liquid does not penetrate any deeper than one particle radius.

The waiting time elapsed prior to the apparition of the liquid cord is indicated in Figure 15 of Tadmor *et al.* [TAD 61]. The screw length corresponding to this waiting time *does not exceed 5%* of the total length of the fusion zone. It is this zone that we will now consider.

5.3.5. *Thermal exchange in the fusion zone*

Tadmor [TAD 66] established the theory for this exchange. He improved his description in 1967 by taking into account that:

– the temperature profile in the melted film is not linear;

– the heat required to heat the polymer from its fusion temperature to that of the film must be accounted for.

Finally, the thickness of the melted film $\overset{o}{\delta}$ is given by equation 24. Equation 25 gives the expression of parameter ϕ that intervenes in the adimensional thermal parameter Ψ:

$$\Psi = \frac{\phi\sqrt{W}}{\left(\frac{X_o}{W}\right)^{\frac{1}{2}}\frac{G}{H_D}}$$

The channel progressively narrows (its depth decreases) according to the law:

$$H = H_o - Az$$

Width X of the solid cord is expressed by author's equation 2:

$$\frac{X}{W} = \frac{X_1}{W}\left[\frac{\Psi}{A} - \left(\frac{\Psi}{A} - 1\right)\left(\frac{H_o}{H_o - AZ}\right)^{\frac{1}{2}}\right]^2$$

Tadmor [TAD 67] also gives the expression of the X/W ratio for a channel of consistent depth.

The experimental results of Tadmor [TAD 67] confirm these calculations.

The thermal power exchanged between the barrel and the product may reach 5 $kW.m^{-2}$ of barrel surface.

5.4. Shaft electrical power

5.4.1. *Shaft power of the screw*

The friction force to overcome between the liquid and the barrel is:

$$F_{fr} = (L_{he} \times 2X)\tau = (L_{he} \times 2X) C V_R^n \quad \text{in the direction of } \vec{V}_R$$

The corresponding power on the shaft is:

$$\Pi_{liq} = \pi N D L_{he} 2XC \left(\frac{2\pi N R \sin\phi}{\sin(\theta + \phi)} \right)^n \cos\theta \quad \text{with } L_{he} = \frac{L}{\sin\phi}$$

Hence, the power on the shaft is:

$$\Pi_T = \Pi_m = \Pi_{liq} \qquad (\text{Watt})$$

Of course, we should account for the motor and transmission yield in order to obtain the electrical consumption.

5.4.2. *Extrapolation of power from diameter d to diameter D*

The torque resistance to overcome varies according to:

$$(ND)^n \times D = N^n D^{1+n}$$

Flowrate is proportional to:

$$ND \times \text{channel cross} - \text{section}$$

We admit that the channel cross-section varies proportionally to D and not D^2.

Therefore, power Π varies per:

$$\Pi \propto N^n D^{1+n} N D^2 = N^{n+1} D^{3+n}$$

We similarly accept that N varies proportionally to $D^{-0.5}$. Hence:

$$\Pi \propto D^{2.5+0.5n}$$

And:

$$\frac{\Pi_D}{\Pi_d} = \left(\frac{D}{d}\right)^{2.5+0.5n}$$

Of course, the above only represents an approximate expression.

5.5. Practical considerations and screw use

5.5.1. *Operating variables and dimensioning variables*

If we increase the rotation velocity of an adiabatic screw, the viscous dissipation increases proportionally to N^2. However, the *viscosity* of the pseudoplastic polymer decreases with the velocity. Furthermore, the flow increases with the rotation velocity and the non-melted proportion can also increase. In practical terms, the rotation velocities remain within the range of 30–120 $rev.mn^{-1}$.

If the barrel temperature increases, the heat transmitted to the polymer by conduction also increases, whereas the product's *viscosity* decreases. It follows that there is an optimal temperature. In any case, there is an advantage in increasing the barrel temperature on screw entry as this decreases the delay in fusion onset.

If we increase the channel depth, the cord velocity decreases, the relative velocity of the barrel and the bed increases, which favors viscous dissipation and subsequently decreases the delay in fusion onset. However, the increased depth exposes less of the solid to the heating surface of the barrel due to the decreased channel width. Here again, there is a compromise to be found. The increased depth increases the production per revolution and the shaft torque.

If the helix angle is decreased, the channel length increases but its width decreases. The contact surface with the barrel remains constant together with the heat transmitted. However, the relative velocity between the bed and the barrel decreases as seen in the velocity triangle in Figure 5.8. The viscous dissipation decreases despite the increased channel velocity.

The clearance between the flight edge and the barrel is significant. Nevertheless, viscous dissipation decreases if this clearance increases with wear on the surfaces in relative movement. However, the greater part of the mechanical power of the shaft is spent by viscous dissipation.

The share of viscous dissipation in the total heat observed in the polymer varies from 40 to 75% according to screw configuration, with the remainder coming from the high temperature of the barrel.

If we need to increase the flowrate, we must increase the length of the fusion zone accordingly.

The flowrate possible for extruder screws varies from 100 to 1,500 $kg.h^{-1}$, with the screw diameter varying from 6 to 20 cm.

5.5.2. *Choosing of certain extruder screws*

Wyman [WYM 75] shows how to calculate both the flowrate and the pressure created by two parallel screws with interweaving flights. These screws turn in the same direction or in opposing directions. Moreover, the distribution of residence times for a pair of screws is less spread than for a single screw. A screw pair can therefore be worthwhile if the components of the paste interact with one another. The residence time is in the order of a minute for a twin screw pair.

An anonymous author [ANO 77] indicated that, for a single screw, going from one flight to two parallel flights increases production by 50%.

For a process using successive loads to fill a mold, Donovan *et al.* [DON 71] measured the evolution of solid proportion along the channel (of a single screw). Screw operation draws all the nearer to the continuous regime as the screw rotation time is a significant share of the cycle.

5.5.3. *Uses of extruders*

These machines are widely used to squeeze grape juice (from grapes separated from the wood beforehand) or in making certain types of oil (from peanuts, cotton, colza, walnuts, grapeseed, etc.). In such cases, the extruder is a continuous press providing both oil and oil cake.

Biological powders undergo both mechanical and thermal action in an extruder. An example of this is cereals puffed by the quick release of heated water present in the product.

The production of noodles, caramels, chewing-gum, marzipan and licorice is also performed with screw extruders.

Seamless metal pipes are obtained by passage through a die; however, extruder screws cannot be used due to the high temperatures.

5.6. Mixing and thermal transfer in the screw

Four authors have studied mixing in screws: Maddock [MAD 59], Pinto and Tadmor [PIN 70], Shimizy [SHI 74] and Sheridan [SHE 75].

Jepson [JEP 53] gives a network of curves that provide the thermal transfer coefficient between a viscous liquid and the barrel surface of the screw. In particular, according to the clearance between the flight and the barrel, we can write:

Clearance (mil)	Coefficient ($Btu.h^{-1}.ft^{-2}.°F^{-1}$)
0	100
10	60
20	35

Remember that:

$$1 \text{ mil} = 25.4 \text{ µm}$$

$$1 \text{ Btu.h}^{-1}.\text{ft}^{-2}.°\text{F}^{-1} = 5.664 \text{ W.m}^{-2}.°\text{C}^{-1}$$

Viscosity varies from 100 to 200 Pa.s. in Jepson's measurements.

6

Choice and Performance of Compressors

6.1. About compressors

6.1.1. *Energy loss and yields*

No compression can be performed without energy dissipation and heat generation. We can discern two types of loss:

– mechanical losses;

– aeraulic aerodynamic losses.

Mechanical losses result from the mutual friction of certain parts of the machine and arise independently from the gas present. The best way to evaluate them is by measuring power when empty, that is, in the absence of gas. This requires the suction and discharge of the machine to be connected to the atmosphere in a continuous manner. If P_M is the shaft power when empty and P_a the shaft power in normal operation, then we define the mechanical yield by:

$$\eta_M = \frac{P_a - P_M}{P_a}$$

Aeraulic losses result directly from the action of the machine on the gas. They are predominant for turbocompressors due to the rotation velocity of between 10 and 100 times greater than that of volumetric compressors, meaning that impacts and friction acting on the gas are more significant. Nonetheless, in a volumetric machine, such as a reciprocating compressor, there are aeraulic losses due to gas laminating in the suction and discharge valves.

Finally, we have:

P_u: useful power compression;

Q_M: thermal power resulting losses;

Q_A: thermal power resulting from aeraulic losses;

P_a: the shaft power of the machine.

We can write:

$$P_a = P_u + Q_M + Q_A$$

Of course, Q_M is identical to the power when empty P_M.

The aerodynamic power of the machine is:

$$P_A = P_u + Q_A = P_a - Q_M$$

The mechanical yield defined above is:

$$\eta_M = \frac{P_a - P_M}{P_a} = \frac{P_A}{P_a}$$

The aerodynamic yield is:

$$\eta_A = \frac{P_u}{P_A}$$

So that the overall yield is:

$$\eta = \frac{P_u}{P_a} = \frac{P_u}{P_A} \times \frac{P_A}{P_a} = \eta_M \eta_A$$

However, a distinction should be made that one of the two yields, aerodynamic or mechanical, is far greater than 1 while the other is close to 1. In this respect, volumetric machines and turbomachines behave very differently.

1) Volumetric compressors have an aerodynamic yield very close to 1 so that the shaft power is simply:

$$P_a = \frac{P_u}{\eta_M}$$

This is how we treat both screw and lobe compressors.

2) On the other hand, turbocompressors have an aeraulic yield far less than 1. Hence, we relate the compression energy VdP to a kilomole of gas and the enthalpy differential of gas is:

$$dH = TdS + VdP = \frac{VdP}{\eta_A}$$

Indeed, according to turbomachine manufacturers, aerodynamic losses are proportional to the compression effort *throughout* compression duration and, consequently, for all elementary steps.

In addition, manufacturers estimate that the mechanical yield of their compressors is close enough to 1 to be integrated with aerodynamic yield, and hence:

$$dH = \frac{VdP}{\eta}$$

η is the overall yield here, known as the polytrophic yield. This word, taken from the Greek, means *multiple directions*. A more appropriate meaning would be *hybrid yield*, due to the mechanical losses not being proportional to the compression power.

The behavior of reciprocating compressors is intermediate between the two extremes above.

6.1.2. *Gas exit temperature*

The accepted hypothesis is that the full shaft power is used to increase gas enthalpy:

$$P_a = \frac{W}{M} C_P (T_1 - T_0)$$

C_P: thermal capacity of the gas (J.kmol^{-1}.°C^{-1}).

Furthermore, we will prove that the thermal loses to the environment are of little importance.

6.1.3. *Energetic losses to the environment*

Since the compressor is hot, it will lose heat by radiation and by convection, particularly, as is always the case, if it is not thermally insulated.

The power lost in kW is expressed by:

$$P_P = \frac{S}{1000}\left[\varepsilon\sigma\left(T_G^4 - T_A^4\right) + \alpha\left(T_G - T_A\right)\right]$$

S: estimate of the machine surface area (m^2);

ε: emissivity of the exchange surface; we take $\varepsilon = 0.8$;

σ: Stefan's constant, $\sigma = 5.67.10^{-8}$ W.m^{-2}.K^{-4};

α: convection coefficient; we take $\sigma = 5$ W.m^{-2}.K^{-1};

T_A: ambient temperature (K);

T_G: gas temperature (K).

T_G is the weighted average of entry (T_0) and exit (T_1) temperatures:

$$T_G = \frac{2T_1 + T_0}{3}$$

EXAMPLE 6.1.–

$$S = 4\ m^2 \quad T_G = 150°C = 423\ K \quad T_A = 20°C = 293\ K$$

$$P_P = \frac{4}{1000}\left[0.8 \times 5.67.10^{-8}\left(423^4 - 193^4\right) + 5\left(423 - 293\right)\right]$$

$$P_P = 0.004\left(1116 + 650\right)$$

$$P_P = 7kW$$

So, a machine with an exterior surface area of 4 m^2 has a power of several hundred kW, which means that these losses will be in the order of 1–2% of the machine's power. Of course, this value can be exceeded if the machine is small in size and operates at a high rate of compression.

These losses correspond to the natural cooling of the machine and approach the compression of an isothermal compression, slightly increasing yield. This effect remains weak and will not be taken into account here.

6.1.4. *Ideal compression power*

Ideal compression refers to the compression of a gas with a machine whose yields are always equal to 1.

To carry a kilomole of perfect gas from pressure P_0 to pressure P_1, useful energy E_u must be applied to the gas, so that:

$$E_u = \int_{P_0}^{P_1} VdP$$

Since the machine is perfect, compression is isentropic. Hence, for a typical gas:

$$dS = C_V \frac{dT}{T} + \left(\frac{\partial P}{\partial T}\right)_V dV$$

$$dS = C_P \frac{dT}{T} - \left(\frac{\partial V}{\partial T}\right)_P dP$$

Make dS = 0 and divide member by member:

$$\frac{C_p}{C_v}\frac{dV}{V} + \frac{dP}{P}\left[\frac{1}{V}\left(\frac{\partial V}{\partial T}\right)_P / \frac{1}{P}\left(\frac{\partial P}{\partial T}\right)_V\right] = 0$$

Therefore, for a perfect gas:

$$\alpha = \frac{1}{V}\left(\frac{\partial V}{\partial T}\right)_P = \beta = \frac{1}{P}\left(\frac{\partial P}{\partial T}\right)_V = \frac{R}{PV}$$

The quantity between the brackets is equal to 1, and if we write:

$$C_P / C_v = \gamma$$

The appendix provides the values of γ for common gases.

We obtain:

$$\gamma\frac{dV}{V} + \frac{dP}{P} = 0 \quad \text{that is: } PV^{\gamma} = \text{const.}$$

Hence:

$$VdP = \frac{P_0^{1/\gamma} V_0}{P^{1/\gamma}} dP$$

And:

$$E_u = \frac{\gamma}{\gamma - 1} P_0^{1/\gamma} V_0 \left[\frac{1}{P_0^{\frac{1}{\gamma}-1}} - \frac{1}{P_1^{\frac{1}{\gamma}-1}}\right]$$

$$E_u = \frac{\gamma}{\gamma - 1} P_0 V_0 \left[\frac{P_1}{P_0}^{\frac{\gamma-1}{\gamma}} - 1\right]$$

$$E_u = \frac{\gamma R T_0}{\gamma - 1}\left[\tau^{\frac{\gamma-1}{\gamma}} - 1\right]$$

And the ideal power P* is:

$$P^* = \frac{W}{M} E_u = \frac{\gamma W}{(\gamma - 1)} \frac{R T_0}{M}\left[\tau^{\frac{\gamma-1}{\gamma}} - 1\right]$$

M and W are, respectively, the molar mass and the flow in mass ($kg.s^{-1}$) of the gas.

6.1.5. *Real gases*

Let us recall the compression equation, which is:

$$dH = \frac{VdP}{\eta}$$

By selecting pressure P as the integration variable, we can express the parallel evolution of:

– temperature;

– spent energy.

Knowledge of a real fluid supposes data concerning:

– its state equation $\quad P = P(V, T)$

– its enthalpy $\quad H = H(V, T)$

Let us differentiate:

$$dP = \frac{\partial P}{\partial V} dV + \frac{\partial P}{\partial T} dT \quad \text{et} \quad dH = \frac{\partial H}{\partial V} dV + \frac{\partial H}{\partial T} dT$$

The first equation gives:

$$dV = \frac{dP}{\partial P/\partial V} - \frac{\partial P/\partial T}{\partial P/\partial V} dT$$

Hence:

$$dH = \frac{\partial H/\partial V}{\partial P/\partial V} dP - \frac{(\partial H/\partial V)(\partial P/\partial T)}{\partial P/\partial V} dT + \frac{\partial H}{\partial T} dT = \frac{VdP}{\eta}$$

The compression evolution equation is consequently:

$$\left[\frac{\partial H/\partial V}{\partial P/\partial V} - \frac{V}{\eta}\right] dP + \left[\frac{\partial H}{\partial T} - \frac{(\partial H/\partial V)(\partial P/\partial T)}{\partial P/\partial V}\right] dT = 0$$

As pressure P varies from P_0 to P_1, we can break the operation into successive steps by giving P increments of ΔP_i. The step of index i is characterized by:

– initially: P_{i-1} and T_{i-1}

– in the end: P_i and T_i

During this step, the mean values are:

$$\overline{P_i} = P_{i+1} + \frac{\Delta P_i}{2}, \quad \overline{V_i} = V_{i-1} + \frac{\Delta V_i}{2} \quad \text{and} \quad \overline{T_i} = T_{i-1} + \frac{\Delta T_i}{2}$$

We will calculate the partial derivatives occurring in the evolution equation according to these mean values. Thus:

$$\left(\frac{\overline{\partial P}}{\partial V}\right)_i \text{ is calculated from } \overline{P_i} \text{ and } \overline{T_i}$$

And the evolution equation e is written, in finite differences:

$$\left[\frac{\overline{(\partial H/\partial V)_i}}{\overline{(\partial P/\partial V)_i}} - \frac{\overline{V_i}}{\eta}\right] \Delta P_i + \left[\frac{\partial H}{\partial T_i} - \frac{\overline{(\partial H/\partial V)_i (\partial P/\partial T)_i}}{\overline{(\partial P/\partial V)_i}}\right] \Delta T_i = 0$$

This equation provides the ΔT_i corresponding to the ΔP_i and, in addition, we have:

$$\Delta V_i = \frac{\Delta P_i}{\left(\partial P/\partial V\right)_i} - \frac{\overline{\left(\partial P/\partial T\right)_i}}{\left(\partial P/\partial V\right)_i}\Delta T_i$$

At the start of the calculation for each step, we can only discern the values $T_{i\text{-}j}$ and $V_{i\text{-}j}$ and not T_i, V_i and, *a fortiori* $\overline{T_i}$ and $\overline{V_i}$. Consequently, we should proceed by iterations with the following as point zero:

$$\overline{V}_i^{(0)} = V_{i-1} \qquad \text{and} \qquad T_i^{(0)} = T_{i-1}$$

The number of iterations necessary for the calculation of each step is a maximum of 3.

Thus, at the end of compression, that is when pressure P reaches value P_1, we also know T_1 and V_1 and can easily deduce: H_1 (V_1, T_1).

Hence, the machine shaft power:

$$P_a = \frac{W}{M}\left(H_1 - H_0\right) \qquad \text{(Watt)}$$

W: gas flowrate ($kg.s^{-1}$)

M: molar mass of gas ($kg.kmol^{-1}$)

H_1 and H_0: molar enthalpy ($J.kmol^{-1}$)

NOTE.–

For volumetric machines, we perform the calculation with $\eta = 1$ and only at the end of this calculation divide by η the shaft power obtained.

6.1.6. *Protecting the machines*

Volumetric reciprocating or non-lubricated screw compressors must be preceded by a filter preventing particles greater than 3 μm with a separation yield of:

– 90% or more for non-lubricated screw machines;

– 98% for reciprocating compressors.

The type of filter medium for effective protection of the compressor has a thickness of 15 mm, creating a pressure drop of 5 mbar near atmospheric pressure and receiving the gas with a frontal velocity in the order of 1 $m.s^{-1}$.

Fans whose blades have a peripheral velocity in the order of 100 $m.s^{-1}$ can tolerate the presence of droplets, but are of course damaged by the arrival of large volumes of liquid.

On the other hand, centrifugal compressors must be protected against vesicular liquids and, if the machine processes vapor, this vapor must be superheated. Indeed, the peripheral velocity of centrifugal compressor wheels can reach 350 $m.s^{-1}$, which demands certain precautions. When a compressor is lubricated, it is typically done with oil and, in order to avoid decomposing this oil, the compressed gas temperature must not exceed 220°C. Furthermore, the gas will be polluted by oil vesicles and any subsequent use of the gas will perhaps require the installation of an oil separator.

The value of lubricated reciprocating or screw compressors stems from their low cost. Moreover, a lubricated screw compressor can accept a gas polluted by particles reaching 20–30 μm.

6.2. Reciprocating compressors

6.2.1. *Use*

Reciprocating compressors are recommended for moderate flowrates (air instruments at 8 bar abs.) or very high pressures:

– 1,000 bars for ammoniac synthesis;

– 2,500 bars for manufacturing low-density polyethylene.

For such pressures, the operation occurs on several stages with intermediate cooling. If we have water at 15 or 20°C, the gases are cooled to a temperature of approximately 35–40°C, and typically to the temperature of the gas arriving on the first level. We should note that a gas heats less in proportion to the extent that its C_P/C_V ratio approaches one.

In the field of high pressures, the gas dissolves in oil, which requires the installation of a system of gas–liquid separators.

The most straightforward means of regulating discharge pressure is the on–off control governed by this pressure. This system is used for compressed air networks.

We can also keep the compressor in continuous operation by keeping the suction valve either open or closed.

6.2.2. *Dead space in reciprocating compressors*

The presence of a dead space means that, at the end of compression, the piston is not moved to the end of the cylinder. In other words, a constant mass M_M of gas always remains in the cylinder. Let us note the energy spent or received by the machine over one cycle.

Dead volume gas decompression, work received:

$$M_M \int_{P_1}^{P_0} PdV - Q_{CP}$$

Q_{CP}: heat resulting from friction of the piston in the cylinder.

Suction, work received:

$$M_A P_0 V_0 - Q_{CA}$$

Q_{CA}: heat released by the laminating of gas through the suction valve;

M_A: mass of suction gas;

P_0: suction pressure;

V_0: specific volume of suction gas.

Compression from $(M_A + M_M)$ of P_0 to P_1 (work done):

$$-(M_A + M_M)\int_{P_0}^{P_1} PdV - Q_{CP}$$

Discharge of mass M_A (work done):

$$M_A P_1 V_1 + Q_{CR}$$

Q_{CR} is the heat released by laminating gas through the discharge valve.

The total work supplied by the machine is:

$$M_A(P_1V_1 - P_0V_0) + Q_{CR} + 2Q_{CP} + Q_{CA} - M_A\int_{P_0}^{P_1} PdV$$

The terms relative to the dead volume have disappeared as compression balances the expansion, and more effectively so, as the reciprocal compressors are equipped with a high-inertia flywheel:

$$\frac{M_A}{\eta}\int_{P_0}^{P_1}[d(PV) - PdV] = \frac{M_A}{\eta}\int_{P_0}^{P_1} VdP$$

This result was expected, but shows that the presence of a dead volume has no influence on the power consumed by the product. The dead volume merely increases the cost due to the cylinder that needs to be oversized.

6.2.3. *Energy and yield*

Mechanical yield can be estimated by the following expression:

$$\eta_M = \frac{1}{1.0906 - 3.53.10^{-5}P^*}$$

Where P* is the power calculated as if the gas were ideal and if the overall yield was equal to 1.

This yield accounts for the various friction forces from mechanical parts.

In addition, the linear velocity of the piston is limited to moderate segment wear:

– if the piston is not lubricated with oil, that is, if it is equipped with graphite or PTFE segments, its velocity must not exceed 2 $m.s^{-1}$ in continuous operation and 2.5 $m.s^{-1}$ in intermittent operation;

– if, on the other hand, the piston is lubricated with oil, its velocity can reach 3.5 $m.s^{-1}$, which allows for a reduced cylinder size.

However, above 2.5 $m.s^{-1}$, the gas laminating by the valves become sensitive, potentially corresponding to an aerodynamic yield of 0.95. If we increase the valve size, this results in an increase in the dead volume, and consequently, increased cylinder size.

Hence, the overall yield is estimated as follows:

Slow piston ($V \ll 2.5\ m.s^{-1}$): $\eta = \eta_M$

Fast piston ($V \gg 2.5\ m.s^{-1}$): $\eta = 0.95\ \eta_M$

EXAMPLE 6.2.–

Compression at 3 bar abs. from 200 $m^3.h^{-1}$ of air taken at 1 bar abs. and 20°C in the first stage of the reciprocal compressor, the piston is considered as slow.

The mass flowrate of air is:

$$W = \frac{200}{3600} \times \frac{273}{293} \times \frac{29}{22.42} = 0.066\ kg.s^{-1}$$

Let us estimate the machine power in the ideal case (for air, $\gamma = 1.4$):

$$P^* = \frac{0.066 \times 1.4 \times 8314 \times 293}{1000 \times 0.40 \times 29} \left[3^{\frac{0.4}{1.4}} - 1 \right]$$

$$P^* = 19.4 \times 0.3687 = 7.15 \text{ kW}$$

The mechanical yield of this stage is evaluated by:

$$\eta_M = \frac{1}{1 + 0.0906 - 3.53.10^{-5} \times 7.15} = 0.917$$

The shaft power for this level is:

$$P_a = 19.4 \times \left[3^{\frac{0.4}{1.4 \times 917}} - 1 \right]$$

$$P_a = 19.4 \times 0.408 = 7.9 \text{ kW}$$

If we had only divided P* by η_M, we would have found:

$$7.15 / 0.917 = 7.8 \text{ kW}$$

This value is somewhat weaker.

The heat capacity of air is:

$$\frac{8314 \times 1.4}{0.4 \times 29} = 1003 \text{ J.kg}^{-1} \text{°C}^{-1}$$

Air heating will be:

$$\frac{7900}{1003 \times 0.066} = 119.3 \text{°C}$$

The air will exit at 139.3°C.

NOTE.–

Membrane compressor:

This type of machine is hermetically sealed so that no leaks can occur:

– between the piston and the cylinder;

– through the membrane stem packing (which is located on the opposite side from the gas relative to the membrane).

Consequently, the diaphragm is well suited to fugitive (ammoniac) or dangerous gases.

The flowrate of these machines is moderate (from 10 to 50 $m^3.h^{-1}$). On the other hand, it can be possible to reach pressures in the order of 1,000 bars using several stages with steel lined with a synthetic material.

6.3. Open volumetric compression – screw compressors and lobe compressors

6.3.1. *Description*

Screw compressors have two parallel screws that gear into one another. Contact between the two screws occurs on a plane perpendicular to the rotation axles and periodically moves along the screws. During rotation, the suction and discharge apertures are periodically on or off to allow either gas movement or compression.

The moving parts of a lobe compressor are two gears with only two teeth each. The form of these teeth differs significantly from those of normal gears. Indeed, both parts have a cross-section shaped like the lemniscate of Bernoulli (except for the central part where the double point is replaced by two rounded forms). The two gears act on one another and ensure gas movement, not between themselves, but in the space between them and the machine case.

6.3.2. *Use*

The acceptable limits for a screw compressor's ratio of compression is in the same order as that of a reciprocal compressor, with the limits being set by the temperature of the compressed gas. Therefore, a refrigerating unit operating at –40°C with a condenser cooled by water at 20°C will require a compression ratio of 12 for C.F.C. 22.

The advantage of screw compressors over reciprocal compressors is in the higher flow rates they can manage (30,000 $m^3.h^{-1}$, for example).

Lobe compressors, such as centrifugal compressors, cannot accept compression ratios below a limit which is in the order of 2, with their suction flowrate typically not in excess of 3,000–5,000 $m^3.h^{-1}$. Compared with centrifugal compressors, lobe compressors have the advantage of avoiding the surge problem.

6.3.3. *Yields*

In open volumetric compression, an estimate (albeit approximate) of the volumetric yield can be obtained by:

$$\eta_V = 0.8 + 0.2\exp\left(-0.03\tau^3\right)$$

In this expression, τ is the compression ratio.

$$\tau = P_1 / P_0$$

When increasing the rotation speed of the machine, the leak flow varies little, while the main flow is more or less proportional to this speed. It follows that the volumetric yield increases. On the other hand, machine wear occurs faster.

The mechanical qualities of the lobe compressor may be evaluated by estimating its power when empty using the expression:

$$P_M = 4.5\ Q_N^2 \qquad \text{(kW)}$$

Q_N is the nominal real suction flowrate ($m^3.s^{-1}$).

The nominal flowrate merely sets the size of the machine in the manufacturer's range and has no bearing on aeraulic losses. Furthermore, this flowrate intervenes with its square, which means that the power when empty is proportional to the square of the rotation velocity.

For screw compressors, we can estimate the mechanical yield by:

$$\eta_M = \frac{1}{1 + 0.00906 - 3.53.10^{-5}.P^*}$$

P^* is the ideal power of gas compression l (ideal gas and $\eta = 1$).

6.3.4. *Internal leak and volumetric yield*

Let us suppose that a compressor is so badly installed that on discharge a leak is produced with flow W_f. To ensure production W_A, the machine must manage the leak flow $W_A + W_f$. Let us write:

$$\eta_v = \frac{W_A}{W_A + W_f}$$

Compared to a machine without a leak, the power required is obtained by dividing by η_V.

Of course, in reality it is quite different. As discharge pressure is greater than suction pressure, counter-current leaks relative to the main flow occur inside the machine. We can model the compressor as represented in Figure 6.1.

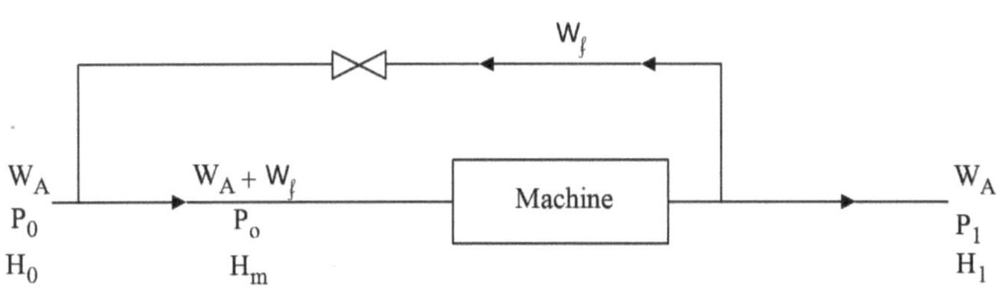

Figure 6.1. *Model of an open volumetric compressor*

The leak flowrate W_f undergoes isenthalpic expansion through an imaginary valve (in reality, this is between the machine rotor and the stator). It expands from P_1 to P_0 and retains the molar enthalpy H_1 corresponding to the compressor discharge.

If we perform an overall balance including the machine and the recirculation loop, we can write out:

$$W_A (H_1 - H_0) = \frac{(W_A + W_f)}{\eta} \int_m^1 V dP$$

This is:

$$H_1 - H_0 = \frac{1}{\eta\,\eta_V}\int_m^1 VdP = \frac{1}{\eta_V \eta} C_P T_m \left(\tau^{\frac{\gamma-1}{\gamma}} - 1 \right)$$

η: polytropic yield announcing mechanical and aeraulic irreversibility

η_V: volumetric yield.

The lower limit of the integral corresponds to state *m* of the mixture, including, on the one hand, the *fresh* supply flowrate W_A, at pressure P_0 and with enthalpy H_0 and, on the other hand, the leak of flowrate W_f, at pressure P_0 and with enthalpy H_1.

As enthalpy is an extensive property, we can write:

$$H_m = \frac{W_A H_0 + W_f H_1}{W_A + W_f} = H_0 \eta_V + H_1 (1 - \eta_V)$$

The behavior of ideal gases is more readily expressed when compared with that of real gases:

1) Ideal gases:

For a machine without any leaks:

$$H_1 - H_0 = \frac{1}{\eta}\int_0^1 VdP = \left[\frac{1}{\eta} \left(\tau^{\frac{\gamma-1}{\gamma}} - 1 \right) \right] C_P T_0 = K^* H_0$$

with:

$$K^* = \left[\frac{1}{\eta} \left(\tau^{\frac{\gamma-1}{\gamma}} - 1 \right) \right] \quad \text{and} \quad \tau = P_1 / P_0$$

but when there is a leak:

$$H_1 - H_0 = \frac{K^* H_m}{\eta_V} = \frac{K^*}{\eta_V} \left[H_0 \eta_V + H_1 (1 - \eta_V) \right]$$

or even:

$$H_1 - H_0 = H_0 K^* + H_1 \left[\frac{1}{\eta_V} - 1 \right] K^*$$

or:

$$H_1 - H_0 = \frac{K^* H_0}{\eta_V - K^* (1 - \eta_V)}$$

The last equality shows that the $H_1 - H_0$ difference is greater than quantity:

$$K^* H_0 / \eta_V$$

Indeed, on expansion, the compression energy of the leak gas is degraded to heat the fresh supply.

The shaft compression power required is:

$$P_a = \frac{W_A K^* H_0}{M \eta \eta_V \left[1 - K^* \left(\frac{1}{\eta_V} - 1 \right) \right]}$$

W_A: supplied flowrate ($kg.s^{-1}$)

M: molar mass of gas ($kg.mol^{-1}$)

H_0 and H_1: molar enthalpy ($J.kmol^{-1}$)

The presence in the denominator of the term $-K^* (1/\eta_V - 1)$ shows that the influence of the volumetric yield increases with K^*, that is, with the compression ratio.

The exit temperature of the gas is obtained by:

$$T_1 = \frac{H_1}{C_P} = \frac{H_0 \left(1 + K^*\right)}{C_P \left[1 - K^* \left(\frac{1}{\eta_V} - 1 \right) \right]} = \frac{T_0 \left(1 + K^*\right)}{1 - K^* \left(\frac{1}{\eta_V} - 1 \right)}$$

Once again with:

$$K^* = \frac{1}{\eta}\left(\tau^{\frac{\gamma-1}{\gamma}} - 1\right)$$

2) Real gases:

Far from critical conditions, it should suffice to correct the results of the accurate calculation using the state equation in the hypothesis where η_V is equal to 1. We then have, marking the calculations without a leak with a cross:

$$T_1 = \frac{T_1^+}{1 - K\left(\frac{1}{\eta_V} - 1\right)}$$

$$P_a = \frac{P_a^+}{\eta_V\left[1 - K\left(\frac{1}{\eta_V} - 1\right)\right]}$$

If we are close to the critical zone, and even more so if we enter this zone, we have to proceed by iterations:

Iteration zero: $H_1^{(0)} - H_0 = \frac{1}{\eta}\int_{P_0}^{P_1} VdP$

As we have seen, this calculation provides T_1 and V_1:

$$H_m^{(0)} = H_0 + \frac{1-\eta_V}{\eta_V} H_1^{(0)}$$

knowing the expression of $H_m = H_m\,(T_m, V_m)$

and that of: $P_0 = P_0\,(T_m, V_m)$

We can deduce T_m and V_m, effectively $P = Cste = P_0$:

$$H_m^{(0)} - H_A = \left[\frac{\partial H}{\partial T} - \frac{(\partial H/\partial V)(\partial P/\partial T)}{\partial P/\partial V}\right]\left(T_m^{(0)} - T_o\right)$$

Hence, $T_m^{(0)} - T_A$. Furthermore:

$$\Delta V = V_m^{(0)} - V = \frac{\partial P/\partial T}{\partial P/\partial V}(T_m - T_0)$$

Hence, V_m.

We can calculate (at iteration 1) $H_1^{(1)} = H_0 + \frac{1}{\eta}\int_{P_m}^{P_1} VdP$

We can then calculate $H_m^{(1)}$, $T_m^{(1)}$, $V_m^{(1)}$, $H_1(2)$ and so forth.

NOTE.–

A different model could accept that the leak flowrate develops progressively through the machine, but here we have taken the most straightforward route.

EXAMPLE 6.3.–

Compression of 5,000 $Nm^3.h^{-1}$ of air from 1 bar abs. to 2.9 bar abs. The air is supplied at 1 bar abs and 20°C. The machine used is a screw compressor.

In kilomoles.s^{-1}, the flowrate supplied is:

$$\frac{5000}{3600 \times 22.42} = 0.0619 \text{ kmol.s}^{-1}$$

The molar thermal capacity is (for air $\gamma = 1.4$):

$$C_P = \frac{8314 \times 1.4}{0.4} = 29099 \text{ J.kmol}^{-1}.\text{K}^{-1}$$

Ideal power:

$$P^* = \frac{0.0619 \times 29099 \times 293}{1000}\left[2.9^{\frac{0.4}{1.4}} - 1\right]$$

$$P^* = 527 \times 0.3555 = 187.36 \text{ kW}$$

Mechanical yield:

$$\eta_M = \frac{1}{1 \times 00906 - 3.53.10^{-8} \times 187.4} = 0.92$$

Volumetric yield:

$$\eta_V = 0.8 + 0.2\exp\left(-0.03 \times 2.9^3\right) = 0.90$$

$$K^* = \frac{1}{0.92}\left[2.9^{\frac{0.4}{1.4}} - 1\right] = 0.3864$$

$$T_1 = \frac{293 \times 1.3864}{1 - 0.3864\left(\dfrac{1}{0.9} - 1\right)} = 424.4 \text{ K} = 151.3°\text{C}$$

$$P_a = \frac{187.36}{0.92 \times 99\left[1 - 0.3864\left(\dfrac{1}{0.9} - 1\right)\right]}$$

$$P_a = 236.4 \text{ kW}$$

EXAMPLE 6.4.–

Air circulation of a dryer with 3,000 $Nm^3.h^{-1}$ at 60°C and 1 bar. The compression required is 1.5. The machine used here is a lobe compressor.

Flow supplied:

$$\frac{3000}{3600\times 22.42}=0.0372\ \text{kmol.s}^{-1}$$

The molar thermal capacity of air is ($\gamma = 1.4$):

$$C_P=\frac{8314\times 1.4}{0.4}=29099\ \text{J.kmol}^{-1}.\text{K}^{-1}$$

Ideal power:

$$P^*=\frac{0.0372\times 29099\times 333}{1000}\left[1.5^{\frac{0.4}{1.4}}-1\right]$$

$$P^*=360.5\times 0.1228=44.3\ \text{kW}$$

Power when empty:

$$P_M=4.5\times\left[\frac{3000}{3600}\times\frac{293}{273}\right]^2=3.6\ \text{kW}$$

Volumetric yield:

$$\eta_V=0.8+0.2\ \exp\left(-0.03\times 1.5^3\right)=0.98$$

$$K^*=\frac{1}{0.917}\left[1.5^{\frac{0.4}{1.4}}-1\right]=0.1339$$

$$T_1=\frac{333\times 1.1339}{1-0.1339\left(\dfrac{1}{0.98}-1\right)}=378.6\ \text{K}=105.5°\text{C}$$

$$P_a=\frac{44.3}{0.98\left[1-0.1339\left(\dfrac{1}{0.98}-1\right)\right]}+3.6$$

$$P_a=48.9\ \text{kW}$$

6.4. Turbo compressors

6.4.1. *Description and use*

Turbocompressors are not, as some believe, compressors moved by a turbine, but are machines in which compression is obtained by kinetic energy and the conversion of this into pressure occurs in a diffuser.

In the chemical industry, we use centrifugal compressors and axial compressors for the most part.

Centrifugal compressors are more flexible than axial machines, with a greater tolerance for supply variations. When equipped with a single wheel, the maximum accepted value for the compression ratio is in the order of 1.8. With two wheels, the economic limit can be in the order of 2.5.

Axial compressors allow for higher rates of compression as they are equipped with a series of stages (wheel groups) separated by intermediate coolers. The rate of compression of one stage is in the order of that of one centrifugal compressor wheel. Axial compressors are sensitive to impurities (erosion, corrosion).

The polytrophic yield of turbomachines typically corresponds to the following ranges:

– one-wheel centrifuge: 0.77–0.83;

– two-wheel centrifuge: 0.74–0.78 (entry into the second wheel is accompanied by irreversibility generating impacts);

– axial compressor level: 0.77–0.83.

Turbocompressors are suited to the following flowrates:

– high: 5,000–15,000 $Nm^3.h^{-1}$ for a centrifugal machine recompressing vapor;

– very high: 50,000 $Nm^3.h^{-1}$ for an axial machine compressing the ethylene of a steam cracker.

The machine's characteristic curve represents the pressure variations on discharge according to the real suction flow. For centrifugal compressors, this characteristic is *flat*, that is, the discharge pressure varies little.

However, when the downstream pressure is too high, the flowrate decreases and, once past its maximum, the pressure created by the machine decreases, which accelerates the reduction in flowrate. The discharge pressure subsequently decreases and flow increases again. These variations are detrimental to compressor operations and are known as *surges*. Surges occur when flow is around half that of nominal flow. It can be corrected by an *anti-surge valve* on discharge, which allows for flow to be kept compressed, either by evacuating the surplus into the atmosphere or by recirculating the surplus back to suction. The valve is controlled by the downstream pressure or by a flow sensor on the compressor feed (before the recycled surplus returns).

6.4.2. *Flowrate regulation*

We can choose between:

– variation in the velocity, since while the flow is proportional to the velocity, the pressure created is proportional to the square of this velocity, which complicates things, making the second solution seem preferable;

– fitting the compressor suction not with a throttling valve but with louvers, which, while they require a small amount of energy, are reliable.

6.4.3. *Energy equation*

Turbocompressors are characterized by their acceleration of the entry gas, conferring it with kinetic energy C. This energy is then converted and absorbed by:

– contributing to the increase in the internal energy of the fluid by compression work (term – PdV);

– providing the energy required for inevitable thermal dissipation (term TdS);

– ensuring the gas transfer (term $P_1V_1 - P_0V_0$).

Hence, from state zero to state 1, in total:

$$-\Delta C = -\int_0^1 PdV + \int_0^1 TdS + \int_0^1 d(PV) = \int_0^1 TdS + \int_0^1 VdP = \Delta H$$

Once again, we meet the principle of conservation of total energy for fluid mechanics:

$$H + C = Cste$$

In the study of turbomachines, we accept that, throughout compression, the term TdS remains proportional to the term VdP so that we write out:

$$dH = C_P dT = \frac{1}{\eta} VdP$$

Indeed, nearly all of the irreversibility is aeraulic in origin.

Now, let us distinguish ideal gases from real gases.

6.4.4. *Ideal gases*

The energy equation is written as:

$$\frac{VdP}{\eta} = dH = C_P dT = \frac{\gamma R}{\gamma - 1} dT = \frac{\gamma}{\gamma - 1}(PdV + VdP)$$

We refer to Appendix Q for the values of γ corresponding to the main industrial gases.

Write out:

$$\frac{\eta\gamma}{\gamma - 1} = \frac{n}{n - 1}$$

We obtain:

$$(n-1)VdP = n(PdV + VdP) \text{ that is: } \frac{dP}{P} + n\frac{dV}{V} = 0$$

Integration gives the law for the so-called polytrophic evolution:

$$PV^n = cste$$

Furthermore, since V = RT/P:

$$dV = \frac{RdT}{P} - \frac{RT}{P^2} dP$$

and:

$$RT\frac{dP}{P} = \frac{n}{n-1}\left(RdT - \frac{RT}{P}dP + \frac{RT}{P}dP \right)$$

$$\frac{dP}{P} = \frac{n}{n-1}\frac{dT}{T} \text{ that is}: \frac{T_1}{T_0} = \frac{P_1}{P_0}^{\frac{n-1}{n}}$$

Thus, we obtain the exit temperature T_1 of gas from the machine.

In addition, we can write:

$$H_1 - H_0 = C_P\left(T_1 - T_0\right) = \frac{\gamma R}{\gamma - 1} T_0 \left[\frac{P_1}{P_0}^{\frac{n-1}{n}} - 1 \right]$$

Machine shaft power is:

$$P_a = \frac{\gamma}{\gamma - 1} \frac{WRT_0}{M} \left[\tau^{\frac{n-1}{n}} - 1 \right]$$

EXAMPLE 6.5.–

Recompression of steam is taken at 100°C and 1.0135 bar abs. with a rate of compression equal to 1.6. The steam flow is 10 tonne.h^{-1}. For steam, γ = 1.33 and we estimate the polytrophic yield at 0.8.

Flow treated:

$$\frac{10000}{18 \times 3600} = 0.154 \text{ kmol.s}^{-1}$$

Molar heat capacity:

$$C_P = \frac{8314 \times 1.33}{0.33} = 33507 \text{ J.kmol}^{-1}.\text{K}^{-1}$$

Temperature of compressed steam:

$$T_1 = 373 \times 1.6^{\frac{0.33}{1.32 \times 0.8}} = 373 \times 1.1568$$

$$T_1 = 431.5 \text{ K} = 158°\text{C}$$

Machine shaft power:

$$P_a = \frac{0.154 \times 33507 \times 373 \times 0.1568}{1000} = 301 \text{ kW}$$

6.4.5. *Real gases*

Since turbocompressors have limited compression ratios, the pressures at work are moderate, which means that we can use the fast method based on the curves of Edmister [EDM 61].

Some manufacturers, by analogy with centrifugal pumps, use the notion of discharge height that has the advantage of not depending (at least in principle) on the process product.

This height is defined by:

$$H = \frac{E_u}{gM}$$

g: acceleration due to gravity (9.81 m.s^{-2})

E_u: useful energy required for compression (J.kmol^{-1})

M: molar mass of gas (kg.kmol^{-1}).

In order to assess the required energy E_u, we need to ignore the energy converted into heat, that is, to perform the calculation in ideal conditions in which the polytropic yield is equal to 1.

Thus, with the data of the previous example:

$$33507 \times 373\left(1.6^{\frac{0.33}{1.33}} - 1\right) = 12.498.10^6 \times 0.1237$$

$$E_u = 1.546.10^6\,\text{J.kmol}^{-1}$$

$$H = \frac{1.546 \times 10^6}{9.81 \times 18} = 8755\ \text{m}$$

Let us examine the de-superheating of compressed steam:

Even if a compressor takes in saturated steam, it discharges superheated steam that needs to be de-superheated in certain applications (for instance in the case of the mechanical compression of steam).

As we see in the tables:

H_0: mass enthalpy of saturated suction vapor (kJ.kg^{-1})

H_1: mass enthalpy of saturated vapor at the discharge pressure (kJ.kg^{-1}).

We calculate the shaft power P_a from which we can deduce the energy received per kilogram of gas:

$$E = \frac{P_a}{W}$$

We can also calculate:

$$h = 4.19\ t_e$$

h and t_e are the mass enthalpy and temperature of de-superheated water. Temperature t_e is measured in degrees Celsius.

We can find the mass of water x to add to 1 kg of steam in order to saturate it, which means that its specific enthalpy becomes equal to H_1. This simply requires us to apply the mixing rule:

$$(1+x)H_1 = x\,h + H_1^{'}$$

Therefore:

$$x = \frac{H_1^{'} - H_1}{H_1 - h}$$

H' is the mass enthalpy of the discharged steam by the compressor at pressure P_1 and at temperature T_1. We can then:

– either read this enthalpy on the Mollier diagram (if we have one) or in the tables;

– or write out:

$$H_1^{'} = H_0 + E$$

The second method is the more correct.

EXAMPLE 6.6.–

$$E = 109 \text{ kJ.kg}^{-1}$$

$H_O = 2{,}676 \text{ kJ.kg}^{-1}$ $\quad H_l = 2{,}696 \text{ kJ.kg}^{-1}$

$T_e = 80°C$ $\quad h = 4.19 \times 80 = 334 \text{ kJ.kg}^{-1}$

Hence:

$$x = \frac{2676 + 109 - 2696}{2696 - 334} = 0.037$$

We must add 3.7% of water to the compressed vapor to de-superheat it.

6.5. Fans

6.5.1. *Use*

The pressure differential of which axial (helix) fans are capable does not exceed 3,500 Pa. However, centrifugal ventilators can provide 15,000 Pa, which is why they are used in combustion installations.

The flowrate treated by ventilators can be high, surpassing 15 $Nm^3.s^{-1}$ close to atmospheric pressure.

Machines with (flat) radial blades are used when the treated gas can lead to solid deposits, as they are resistant to fouling.

6.5.2. *Compression power*

As the over-pressurization created by these machines is slow, the corresponding compression ratio varies very little from 1 and can be written by a development of the first order:

$$\tau^{\frac{n-1}{n}} = e^{\frac{n-1}{n}\text{Ln}\tau} \# 1 + \frac{n-1}{n}\text{Ln}\tau = 1 + \frac{\gamma-1}{\gamma\eta}\text{Ln}\tau$$

and for the machine shaft power:

$$P_a = N\frac{R\gamma}{\gamma-1}T_0\left[\tau^{\frac{n-1}{n}} - 1\right] \# \frac{NRT_0}{\eta}\text{Ln}\frac{P_1}{P_0}$$

This expression is exactly that of an isothermal compression with energetic yield η.

In reality, compression is clearly not isothermal and we have:

$$T_1 = T_0\tau^{\frac{n-1}{n}} = T_0\left[1 + \frac{(\gamma-1)}{\gamma\eta}\text{Ln}\tau\right]$$

Fans do not consume much energy relative to the flow treated, and for an estimate it should suffice to use:

$\eta = 0.7$ (and $\eta = 0.5$ for flat bladed machines).

EXAMPLE 6.7.–

For an air ventilator:

$P_0 = 1.135$ bar $\qquad P_1 - P_0 = 10{,}000$ Pa

$T_0 = 293$ K $\qquad W = 5\ \text{kg.s}^{-1}$

$\gamma = 1.4$

Molar flowrate:

$$N = \frac{W}{M} = \frac{5}{29} = 0.172\ \text{kmol.s}^{-1}$$

$$P_a = \frac{0.172 \times 8314 \times 293}{1000 \times 0.7} \text{Ln} \frac{111350}{101350}$$

$$P_a = 598.56 \times 0.094 = 56.26\ \text{kW}$$

The compression ratio is:

$$\tau = 111350 / 101350 = 1.0987$$

The discharge temperature is:

$$T_1 = 293 \times 1.0987^{\frac{0.4}{1.4}} = 304.5\ \text{K} = 31.3°\text{C}$$

which is a heat increase of 11.3°C.

NOTE.–

Let us write, for the pressure:

$$\Delta P = P_1 - P_0$$

The shaft power becomes:

$$P_a = \frac{NRT_0}{\eta} Ln\left[\frac{P_0 + \Delta P}{P_0}\right] \# \frac{NRT_0 \Delta P}{\eta\, P_0}$$

With the previous data, we would have found:

$$P_a = 598.56 \times \frac{10000}{101350} = 598.56 \times 0.0987$$

$$P_a = 59.06 \text{ kW}$$

or 5% error by excess.

6.6. Liquid ring pumps

6.6.1. *Principle and use*

These pumps are widely used as vacuum pumps. Their working principle is close to that of the paddle pumps used in laboratories, but the seal between the paddles and the pump casing is ensured by a peripheral liquid ring partially penetrating the sectors defined by the paddles and held in place by the centrifugal force on the elliptical casing. The high frequency of the paddles in front of to the discharge aperture ensures a regular flow.

The liquid is often water as the suction mixture is frequently a mixture of inert gas and some steam that condenses on contact with the cold water. However, we can typically:

– either use a liquid of the same nature as the condensable vapor;

– or use a liquid that is non miscible with the vapor condensate and separate the two liquids by decantation.

The suction gas flow cannot exceed 6,000–7,000 $m^3.h^{-1}$.

6.6.2. *Shaft power*

Liquid ring pumps are a particular type of volumetric compressors. In addition, since the suction gas is under low pressure (less than atmospheric pressure), we can legitimately consider the suction mixture as an ideal gas.

If Q is the suction flow in real m^3 (in the suction conditions, P_0 mbar and T_0 K) per hour, the flow N in kmoles per second is:

$$N = \frac{Q}{3600 \times 22.42} \times \frac{P_0}{1013} \times \frac{273}{T_0} \qquad \left(\text{kmol.s}^{-1}\right)$$

If τ is the compression ratio, the unit energy expended per kilomole is:

$$E_u = \frac{RT_0}{1000} \frac{\gamma}{\gamma - 1} \left(\tau^{\frac{\gamma-1}{\gamma}} - 1 \right)$$

These machines systematically discharge into the atmosphere, so clearly we have:

$$\tau = \frac{1013}{P_0} (P_0 \text{ in mbar abs.})$$

Useful power is:

$$P_u = \frac{W}{M} E_u$$

We can estimate shaft power by the simple relationship:

$$P_a = \frac{P_u}{\eta}$$

with:

$$\eta = 0.12 + 0.33 \text{ Ln}\tau - 0.1 (\text{Ln}\tau)^2$$

EXAMPLE 6.8.–

$Q = 255 \text{ m}^3.\text{h}^{-1}$ $\qquad P_0 = 80$ mbar

$\gamma = 1.4$ (air) $\qquad T_0 = 20°\text{C} = 293$ K

$$N = \frac{255}{3600 \times 22.42} \times \frac{80}{1013} \times \frac{273}{293} = 2.33.10^{-4} \text{kmol.s}^{-1}$$

$$\tau = 1013/80 = 12.7$$

$$E_u = \frac{8314 \times 293}{1000} \times \frac{1.4}{1.4-1}\left[12.7^{\frac{0.4}{1.4}} - 1\right] = 9111$$

$$P_u = 2.33.10^{-4} \times 9111 = 2.12 \text{ kW}$$

$$\eta = 0.12 + 0.33 \text{ Ln}(12.7) - 0.1 \text{ Ln}(12.7)^2 = 0.33$$

$$P_a = \frac{2.12}{0.33} = 6.5 \text{ kW}$$

6.6.3. *Liquid consumption*

The liquid ring pumps are typically conceived to operate with a ring temperature less than or equal to 18°C. Hence, the ring tends to heat up for two reasons:

– the majority of the vapor present condenses;

– the gas is partially cooled by the liquid ring.

The consequence of the latter observation is that the adiabatic hypothesis for the compression of an ideal gas is pessimistic. In reality, compression follows an intermediate path between adiabatic and isothermic.

In any case, we accept that the thermal power to evacuate from the liquid is:

$$q = 0.9 \ P_a + LW_v \qquad \text{(kW)}$$

P_a: shaft power (kW)

W_v: condensable flowrate of suction vapor ($kg.s^{-1}$)

L: latent condensation heat ($kJ.kg^{-1}$).

Cooling can occur in two ways:

1) auxiliary exchanger:

– the water flows in a circle, crossing a small plate exchanger. If t_R is the temperature of the available cooling water, we calculate the flow required for it to heat up by 2°C. The ring water itself must be cooled by 3°C (from 18 to 15°C). Exchanger power will clearly be q. This method is used when the cooling water cannot be introduced directly into the pump;

2) lost water:

inversely, if the water can be introduced to the pump, its flow will be:

$$Q_E = \frac{q \times 3600}{4180 \times (18 -) t_R} \qquad (m^3.h^{-1})$$

7

Free Gas Expansion

7.1. Types of expansion: one-dimensional flow equations

7.1.1. *Types of expansion and starting hypotheses*

We can distinguish two types of gas expansion, in which a gas either performs work or does not perform work. We will exclude cases where the gas works, for instance on going through a turbine.

Expansion without work is a *free* expansion. It is this type of expansion that occurs in equipment as varied as control valves, safety valves and gas pipelines.

Through this chapter, we will treat the gas as an ideal gas. Indeed, energetic balances are of little significance here, since the energy is lost in any case. Moreover, pressure drop calculations are particularly difficult in the case of safety valves, but they are required from the low-pressure side, where the behavior of the gas is more like that of an ideal gas.

7.1.2. *Free expansion and energy*

When a compressible and viscous fluid goes through a restriction, its velocity is highest at the neck of this restriction. Indeed, the path cross-section is minimum at this point and the flow conservation is:

$$W = \rho u A$$

A: path cross-section (m^2)

u: fluid velocity ($m.s^{-1}$)

ρ: fluid density ($kg.m^{-3}$)

W: flowrate ($kg.s^{-1}$)

Density varies slightly, and the velocity is approximately inversely proportional to A.

After the neck, the cross-section increases again. Depending on the value of the velocity at the neck, three types of phenomenon can occur:

1) Very moderate velocity (Reynolds number less than 2,000).

Only the viscous friction on the wall is involved. Clearly, viscosity is the determining parameter here and the flow occurs in laminar regime. The pressure drop in the restriction is so low that the gas expansion almost never occurs at this location.

2) Significant velocity (Reynolds number more than 5,000).

Vortices in the divergent lead to the greater part of the kinetic energy of the gas being dissipated on neck exit. This is how gas flow control valves work, which we will examine below.

3) High velocity (more than 200 $m.s^{-1}$)

The velocity of sound is reached at the neck and somewhere downstream a stationary shockwave develops, which corresponds to a degradation of acoustic energy into heat.

Indeed, we know that when a solid (airplane) moves relative to a gas, it provokes disturbances in the latter in the form of sound waves. Inversely, a gas that moves relative to a solid surface is similarly subject to sound waves. When the expansion ratio of the gas is barely more than 1, the sound energy created is negligible. On the other hand, when the expansion rate, and consequently the neck velocity, increases, the sound energy is dissipated by a stationary shock wave, and this energy is common in the release of numerous types of security valves as we will see in an example below.

Irrespective of the expansion regime, when a fluid neither receives work nor heat from its environment, its total energy remains constant. Neglecting the macroscopic potential energy, that is, gravity, the total energy of a flowing fluid is the total of its enthalpy H and its macroscopic kinetic energy C:

$$H + C = \text{const.}$$

Finally, the mechanism of all expansions is as follows:

1) acquisition of kinetic energy to the detriment of the enthalpy up to the neck of the restriction;

2) dissipation of heat as part of the kinetic energy acquired in the neck by mechanisms of viscous friction, vortices or a shock wave occurring after the neck (term TdS of the enthalpy differential);

3) transformation of the rest of the kinetic energy into pressure energy in the divergent (term VdP).

Steps 2 and 3 correspond to an increase in the enthalpy to the detriment of kinetic energy. Mechanism 2 concerns heat (TdS) and mechanism 3, pressure (VdP).

In practice, the conservation of total energy is written differently for liquids and for gases:

1) Liquids:

$$H + C = U + PV + C = U + P\frac{M}{\rho} + M\frac{u^2}{2} = \text{Cste}$$

M is the molar mass of liquid and ρ its density. Enthalpy is taken for a kilomole, but as we take the internal energy of 1 kg of fluid, we obtain the practical expression:

$$E + \frac{P}{\rho} + \frac{u^2}{2} = \text{const.}$$

With:

$$E = U/M$$

Expansion then corresponds to a singular pressure drop ΔP, so that:

$$\frac{\Delta P}{\rho} = \xi \frac{u^2}{2}$$

Increase ΔE of the internal energy to the detriment of enthalpy circulation PV translates to an increase in heat.

2) Perfect gases:

The enthalpy of perfect gases depends only on temperature and we write out:

$$H = C_P T + \text{const.}$$

But:

$$C_P - C_V = R \quad \text{and} \quad C_P / C_V = \gamma$$

In the appendix are the values of γ for various gases. Here, we will use the results of Brun *et al.* [BRU 68].

Hence:

$$C_P = \frac{\gamma R}{\gamma - 1} \quad \text{and} \quad H = \frac{\gamma RT}{\gamma - 1} + \text{const.}$$

On the other hand, the state equation for perfect gases is:

$$\frac{RT}{M} = \frac{P}{\rho} \quad \text{that is} \quad H = \frac{M\gamma}{\gamma - 1} \frac{P}{\rho} + \text{const.}$$

M is the molar mass of gas and ρ its density.

The conservation of total energy is then written as:

$$\frac{H + C}{M} = \frac{\gamma P}{(\gamma - 1)\rho} + \frac{u^2}{2} = \text{const.}$$

7.1.3. *Mach number*

The local Mach number M is defined as follows:

$$M^2 = \frac{\rho u^2}{\gamma P} = \frac{u^2}{(\gamma - 1) C_P T}$$

This equation expresses that the Mach number M is equal to the ratio of local velocity u of the fluid to the speed of sound at the same location.

To ease this writing, we will first write out the following:

$$Y = M^2$$

This notation is very important here.

NOTE (Joule–Thomson expansion).–

Let us suppose that, in a laboratory experiment, the velocity of gas is very low, both before and after the restriction. The variation in its kinetic energy can then be ignored and the energy equation shows that the expansion can be considered as isenthalpic. The dissipation of pressure energy only occurs due to viscous friction in laminar regime.

Unlike that of ideal gases, the enthalpy of real gases depends not only on the temperature but also on the pressure and, according to upstream conditions, the enthalpy can be an increasing or decreasing function of pressure, while it is always an increasing function of the temperature. It follows that if enthalpy must remain constant, the pressure drop effect has to be compensated by a variation in the temperature, so that following an isenthalpic expansion, real gases are susceptible to cool or to heat.

Below, we will consider an example from an industrial point of view.

7.1.4. *Pipe friction*

We will render friction using the relationship:

$$-dP = f \frac{\rho u^2}{2} \frac{dL}{D}$$

dL and D: basic length and diameter of the pipe (m)

ρ: density of gas ($kg.m^{-3}$)

u: gas velocity ($m.s^{-1}$)

f: friction coefficient.

Coefficient f is the function of the Reynolds number and is written as:

$$Re = \frac{uD\rho}{\mu} = \frac{4W}{\mu P}$$

μ: gas viscosity: (Pa.s)

W: mass-flowrate of gas ($kg.s^{-1}$)

P: wetted perimeter ($P = \pi D$).

We observe that Re and, consequently f, depends only on pressure and temperature by the intermediary of gas viscosity. Within the ranges of practical variation, this dependence is slow and occurs in turbulent regime. Here, we accept that:

$$f \neq 0.03$$

In practice, we must often use the existing empirical relationships between f and Re (see [BRU 68]).

In the flow equations, we write out:

$$\frac{fdL}{2D} = dX$$

7.1.5. *General equations of adiabatic flow*

This is the flow of an ideal gas with viscous friction and variation in the path cross-section, but without heat exchange with the exterior, which makes the flow adiabatic.

From the four fundamental equations:

– state equation;

– flow conservation (continuity equation);

– energy;

– impulse (momentum),

we obtain, by linear combinations:

$$\frac{1}{2}(\gamma-1)\frac{dY}{Y}=\left[1+\left(\frac{\gamma-1}{2}\right)Y\right]\frac{dA}{A}-\gamma Y\left[1+\left(\frac{\gamma-1}{2}\right)Y\right]dX$$

$$(\gamma-1)\frac{dP}{P}=-\gamma Y\frac{dA}{A}+\left[1+(\gamma-1)Y\right]\gamma\ YdX$$

$$(Y-1)\frac{du}{u}=\frac{dA}{A}-\gamma\ YdX$$

$$(\gamma-1)\frac{d\rho}{\rho}=-Y\frac{dA}{A}+\gamma\ YdX$$

Even if we see a relationship between path cross-section A and the friction parameter X, a direct analytical integration of these equations does not seem possible, meaning that a numerical method is required, which avoids such problems.

On the other hand, if we must independently vary A and X, integration is possible and we will provide the result of the calculation for:

– flow without friction for variable cross-section;

– flow with friction for constant cross-section.

Furthermore, we will use the four fundamental equations in a finite form for the study of:

– the pure shock wave;

– the singular pressure drop.

In both cases, the cross-section variation does not intervene, since the phenomenon studied here is supposed to occur on a plane of zero thickness and perpendicular to the flow.

7.1.6. *Flow without friction but with variable cross-section*

As there is no energy degradation, this flow is isentropic. The equations for the problem are as follows:

Impulse:

$$\frac{dP}{P}+\gamma Y\frac{du}{u}=0$$

Energy:

$$\frac{dT}{T}+\frac{(\gamma-1)}{\gamma}Y\frac{du}{u}=0$$

Continuity:

$$\frac{d\rho}{\rho}+\frac{du}{u}+\frac{dA}{A}=0$$

State equation:

$$\frac{dP}{P}-\frac{d\rho}{\rho}-\frac{dT}{T}=0$$

By linear combination member to member with the four previous equations, we obtain A as an independent variable (A is the cross-section open to the fluid process).

1) Mach number equation (Y equation):

$$\frac{dA}{A}=\frac{1}{2}\frac{(Y-1)dY}{Y\left(1+\frac{\gamma-1}{2}Y\right)}=\frac{(Y-1)dY}{(\gamma-1)Y\left(Y+\frac{2}{\gamma-1}\right)}$$

$$2\frac{dA}{A}=\frac{dY}{(\gamma-1)}\left[\frac{\gamma+1}{Y+\dfrac{2}{\gamma-1}}-\frac{\gamma-1}{Y}\right]$$

That is:

$$\left[\frac{A_2}{A_1}\right]^2=\left[\frac{Y_2+\dfrac{2}{\gamma-1}}{Y_1+\dfrac{2}{\gamma-1}}\right]^{\frac{\gamma+1}{\gamma-1}}\times\frac{Y_1}{Y_2}$$

We can put this equation in the following two forms, for an iterative calculation of Y_2:

$$Y_2^{(n+1)}=\left(Y_1+\frac{2}{\gamma\text{-}1}\right)\left[\frac{Y_2^{(n)}}{Y_1}\left(\frac{A_2}{A_1}\right)^2\right]^{\frac{\gamma\text{-}1}{\gamma\text{-}1}}-\frac{2}{\gamma\text{-}1} \qquad [7.1]$$

$$Y_2^{(n+1)}=Y_1\left[\frac{Y_2^{(n)}+\dfrac{2}{\gamma-1}}{Y_1+\dfrac{2}{\gamma-1}}\right]^{\frac{\gamma+1}{\gamma+1}}\left(\frac{A_1}{A_2}\right)^2 \qquad [7.2]$$

When flow is sonic at the neck:

– form [7.1] allows for the iterative calculation of Y_2 corresponding to the supersonic regime in the divergent;

– form [7.2] corresponds to the subsonic regime.

In both cases, we can take:

$$Y_2^{(0)}=Y_1$$

2) Pressure equation:

$$\frac{dP}{P}=\frac{-\gamma Y}{(Y-1)}\frac{dA}{A}$$

By dividing member to member by the Mach number equation:

$$\frac{dP}{P} = \frac{-\gamma dY}{(\gamma-1)\left(Y + \dfrac{2}{\gamma+1}\right)}$$

$$\frac{P_2}{P_1} = \left[\frac{Y_1 + \dfrac{2}{\gamma-1}}{Y_2 + \dfrac{2}{\gamma-1}}\right]^{\frac{\gamma}{\gamma-1}}$$

Inversely, we can write:

$$Y_2 = \left[\frac{P_1}{P_2}\right]^{\frac{\gamma-1}{\gamma}} \left(Y_1 + \frac{2}{\gamma-1}\right) - \frac{2}{\gamma-1}$$

3) Velocity equation:

$$\frac{du}{u} = \frac{1}{Y-1}\frac{dA}{A}$$

By division with the Mach number equation:

$$\frac{du}{u} = \frac{dY}{(\gamma-1)Y\left(Y + \dfrac{2}{\gamma-1}\right)}$$

$$\frac{du}{u} = \frac{dY}{2}\left[\frac{1}{Y} - \frac{1}{Y + \dfrac{2}{\gamma-1}}\right]$$

$$\frac{u_2}{u_1} = \left[\frac{1 + \dfrac{2}{(\gamma-1)Y_1}}{1 + \dfrac{2}{(\gamma-1)Y_2}}\right]^{1/2}$$

4) Density and temperature equations:

The continuity equation directly gives:

$$\rho_2 = \rho_1 \frac{u_1 A_1}{u_2 A_2} = \frac{W}{u_2 A_2}$$

The perfect gas state equation gives:

$$T_2 = T_1 \times \frac{\rho_1}{P_1} \times \frac{P_2}{\rho_2} = \frac{MP_2}{R\rho_2}$$

M: molar mass (kg.kmol^{-1})

R: ideal gas constant (8,314 J.kmol^{-1}.K^{-1}).

7.1.7. *Isentropic flow and critical values*

Flow can be considered as isentropic in control valve and safety valve convergers. Indeed, the length of these convergers is only several centimeters and friction with the wall can therefore be disregarded.

As a general rule, we know that:

$$dH = TdS + VdP$$

$$dH = C_P dT - \left[T\left(\frac{\partial V}{\partial T}\right)_P - V \right] dP$$

If entropy is constant:

$$dH = VdP$$

If the gas is ideal:

$$dH = C_P dT = \frac{\gamma R}{\gamma - 1} dT$$

But:

$$V = RT/P \qquad \text{so:} \qquad RT\frac{dP}{P} = \frac{\gamma R}{\gamma - 1}dT$$

Let us integrate:

$$\frac{T_1}{T_0} = \left[\frac{P_1}{P_0}\right]^{\frac{\gamma-1}{\gamma}}$$

If flow is critical at the plug of the safety valve, the equation of flow relative to pressure is written (with $Y_c = 1$ and $Y_o = 0$) as:

$$\frac{P_c}{P_o} = \left[\frac{0 + \frac{2}{\gamma - 1}}{1 + \frac{2}{\gamma - 1}}\right]^{\frac{\gamma}{\gamma-1}} = \left(\frac{2}{\gamma + 1}\right)^{\frac{\gamma}{\gamma-1}}$$

$$\text{Et}: \frac{T_c}{T_o} = \left[\left(\frac{2}{\gamma + 1}\right)^{\frac{\gamma}{\gamma-1}}\right]^{\frac{\gamma-1}{\gamma}} = \frac{2}{\gamma + 1}$$

And of course:

$$\rho_c = \frac{MP_c}{RT_c}$$

Velocity at the plug is equal to the speed of sound:

$$u_c = \sqrt{\frac{\gamma P_c}{\rho_c}}$$

7.1.8. *Flow at constant cross-section with friction (gas pipelines)*

1) Mach number equation (equation in Y):

$$\frac{dY}{Y}=\frac{2\gamma Y\left(1+\left(\frac{\gamma-1}{2}\right)Y\right)}{Y-1}dX$$

That is:

$$dX=\frac{(1\text{-}Y)\,dY}{\gamma(\gamma\text{-}1)Y^2\left(Y+\frac{2}{\gamma\text{-}1}\right)} \qquad [7.3]$$

$$dX=\frac{1}{2\gamma}\left[\frac{1}{Y^2}-\frac{(\gamma+1)}{2Y}+\frac{\gamma+1}{2\left(Y+\frac{2}{\gamma+1}\right)}\right]dY$$

And integrating:

$$X_1-X_0=\frac{1}{2\gamma}\left[\left(\frac{\gamma+1}{2}\right)\text{Ln}\left(1+\frac{2}{Y(\gamma-1)}\right)-\frac{1}{Y}\right]_0^1$$

The problem at present is in calculating Y_2 knowing Y_1 and the pipe length, that is, $X_1 - X_0$. For this, we can proceed by iterations, with:

$$Y_2^{(n+1)}=2\gamma(X_1-X_2)+\frac{1}{Y_1}+\text{Ln}\left[\frac{1+\frac{2}{Y_2^{(n)}(\gamma-1)}}{1+\frac{2}{Y_1(\gamma-1)}}\right]^{\frac{\gamma+1}{2}}$$

We choose as an initial value:

$$Y_2^{(0)}=Y_1$$

2) Pressure equation:

$$\frac{dP}{P}=\frac{\gamma Y\left(1+(\gamma\text{-}1)Y\right)}{Y\text{-}1}dX \qquad [7.4]$$

By expression-to-expression division of [7.3] by [7.4], we get:

$$\frac{dP}{P}=-\frac{\left(Y+\frac{1}{\gamma-1}\right)}{Y\left(Y+\frac{2}{\gamma-1}\right)}dY$$

$$\frac{dP}{P}=-\frac{1}{2}\left[\frac{1}{Y}+\frac{1}{Y+\frac{2}{\gamma-1}}\right]$$

And by integration:

$$P_2=P_1\left[\frac{Y_1\left(Y_1+\frac{2}{\gamma-1}\right)}{Y_2\left(Y_2+\frac{2}{\gamma-1}\right)}\right]^{1/2}$$

Inversely, if we want to calculate Y_2 according to P_2, we should be able to resolve the second-degree equation:

$$Y^2+\frac{2Y}{\gamma-1}-\left(\frac{P_1}{P_2}\right)^2 Y_1\left(Y_1+\frac{2}{\gamma-1}\right)=0$$

The physical root of which is:

$$Y_2=\frac{-1}{\gamma-1}+\sqrt{\left(\frac{1}{\gamma-1}\right)^2+\left(\frac{P_1}{P_2}\right)^2 Y_1\left(Y_1+\frac{2}{\gamma-1}\right)}$$

3) Velocity equation:

$$\frac{du}{u} = \frac{-\gamma Y}{Y-1} dX$$

By dividing member to member by [7.3]:

$$\frac{du}{u} = \frac{dY}{(\gamma-1)Y\left(Y+\frac{2}{\gamma-1}\right)} = \frac{1}{2}\left[\frac{1}{Y} - \frac{1}{Y+\frac{2}{\gamma-1}}\right]$$

By integration:

$$u_2 = u_1 \left[\frac{Y_2\left(Y_1+\frac{2}{\gamma-1}\right)}{Y_1\left(Y_2+\frac{2}{\gamma-1}\right)}\right]^{1/2}$$

4) Density and temperature equations:

The continuity equation directly gives:

$$\rho_2 = \rho_1 \frac{u_1}{u_2}$$

The perfect gas state equation also readily gives us:

$$T_2 = T_1 \frac{\rho_1}{P_1} \times \frac{P_2}{\rho_2} = \frac{MP_2}{R\rho_2} \qquad \text{(M: molar mass)}$$

We must verify that:

$$Y_2 = \frac{\rho_2 u_2^2}{\gamma P_2}$$

7.1.9. *Equation of the stationary shock wave*

On the surface unit of the shock wave, the pressure $P_1 - P_2$ throughout the wave creates the increase in the quantity of flow movement:

$$u_2^2\rho_2 - \rho_1 u_1^2$$

So, the impulse equation is written in the finite form as:

$$P_1 + \rho_1 u_1^2 = P_2 + \rho_2 u_2^2 \qquad [7.5]$$

The continuity equation is:

$$\rho_1 u_1 = \rho_2 u_2 \qquad [7.6]$$

The energy equation is, for ideal gases:

$$u_1^2 + \frac{2\gamma}{\gamma-1}\frac{P_1}{\rho_1} = u_2^2 + \frac{2\gamma}{\gamma-1}\frac{P_2}{\rho_2} = \text{cste} = \frac{\gamma+1}{\gamma-1}a^{*2}$$

For the sake of convenience, we write out, by definition of a^*:

$$u_1^2 + \frac{2\gamma}{\gamma\text{-}1}\frac{P_1}{\rho_1} = \frac{\gamma+1}{\gamma\text{-}1}a^{*2} \qquad [7.7]$$

And:

$$u_2^2 + \frac{2\gamma}{\gamma\text{-}1}\frac{P_2}{\rho_2} = \frac{\gamma+1}{\gamma\text{-}1}a^{*2} \qquad [7.8]$$

Let us divide [7.5] by [7.6] member to members:

$$\frac{1}{u_1}\frac{P_1}{\rho_1} + u_1 = \frac{1}{u_2}\frac{P_2}{\rho_2} + u_2 \qquad [7.9]$$

Let us carry the values of P_1/ρ_1 and P_2/ρ_2 taken from [7.7] and [7.8] in [7.9]:

$$u_1 - \frac{(\gamma-1)}{2\gamma}u_1 + \frac{\gamma+1}{2\gamma}\frac{a^{*2}}{u_1} = u_2 - \frac{(\gamma-1)}{2\gamma}u_2 + \frac{\gamma+1}{2\gamma}\frac{a^{*2}}{u_2}$$

Hence:

$$\left(\frac{\gamma+1}{2\gamma}\right)(u_2-u_1)\left(1-\frac{a^{*2}}{u_1u_2}\right)=0$$

The solution of this equation corresponds to the Prandtl equation:

$$u_1u_2=a^{*2}$$

Now, we introduce the Mach number, recalling that:

$$(\text{Mach number})^2=\rho u^2/\gamma\,P=Y$$

The equations are then written as:

impulse $$P\left(Y+\frac{1}{\gamma}\right)=\text{const.} \qquad [7.10]$$

continuity $$\frac{P}{u}Y=\text{const.} \qquad [7.11]$$

Dividing expression to expression [7.10] by [7.11]:

$$u\left(1+\frac{1}{\gamma\,Y}\right)=\text{const.} \qquad [7.12]$$

The procedure required can be as follows:

– calculation of u_2 by the Prandtl equation.

– calculation of Y_2 by equation [7.12].

– calculation of P_2 by equation [7.10].

– calculation of ρ_2 by the continuity equation:

$$\rho_2=\rho_1u_1/u_2$$

– calculation of T_2 by the state equation:

$$\frac{\rho T}{P}=\text{cste that is: }T_2=T_1\frac{P_2\rho_1}{P_1\rho_2}=\frac{MP_2}{R\rho_2}$$

We must verify that:

$$Y=\frac{\rho_2 u_2^2}{\gamma P_2}$$

7.1.10. *Singular pressure drop*

In fact, this term is ill-suited; a more accurate term would be the loss of singular impulse. Indeed, the impulse equation is:

$$\left(P_1+\rho_1 u_1^2\right)-\left(P_2+\rho_2 u_2^2\right)=\xi\rho_1 u_1^2$$

The continuity equation is:

$$\rho_1 u_1=\rho_2 u_2=W \qquad [7.13]$$

The energy equation is:

$$\frac{u_1^2}{2}+\frac{\gamma}{\gamma-1}\frac{P_1}{\rho_1}=\frac{u_2^2}{2}+\frac{\gamma}{\gamma-1}\frac{P_2}{\rho_2}$$

The state equation acts to determine temperature T_2.

We can write out the energy equation:

$$u_1^2+\frac{2\gamma}{\gamma\text{-}1}\frac{P_1}{\rho_1}=\frac{\gamma+1}{\gamma\text{-}1}a^{*2} \qquad [7.14]$$

$$u_2^2+\frac{2\gamma}{\gamma\text{-}1}\frac{P_2}{\rho_2}=\frac{\gamma+1}{\gamma\text{-}1}a^{*2} \qquad [7.15]$$

And the impulse equation:

$$P_1+(1\text{-}\xi)\rho_1 u_1^2=P_2+\rho_2 u_2^2 \qquad [7.16]$$

Dividing member to member [7.16] by [7.13]:

$$\frac{1}{u_1}\frac{P_1}{\rho_1}+u_1(1\text{-}\xi)=\frac{1}{u_2}\frac{P_2}{\rho_2}+u_2 \qquad [7.17]$$

The energy equations give:

$$\frac{P_1}{\rho_1}=\frac{\gamma+1}{2\gamma}a^{*2}-\frac{\gamma-1}{2\gamma}u_1^2$$

$$\frac{P_2}{\rho_2}=\frac{\gamma+1}{2\gamma}a^{*2}-\frac{\gamma-1}{2\gamma}u_2^2$$

Let us write these values in [7.17]:

$$\frac{1}{u_1}\left(\frac{\gamma+1}{2\gamma}a^{*2}-\frac{\gamma-1}{2\gamma}u_1^2\right)+u_1(1-\xi)=\frac{1}{u_2}\left(\frac{\gamma+1}{2\gamma}a^{*2}-\frac{\gamma-1}{2\gamma}u_2^2\right)+u_2$$

$$\frac{\gamma+1}{2\gamma}\frac{a^{*2}}{u_1}+u_1\left[1-\xi-\left(\frac{\gamma-1}{2\gamma}\right)\right]=\frac{\gamma+1}{2\gamma}\frac{a^{*2}}{u_2}+u_2\left[1-\left(\frac{\gamma-1}{2\gamma}\right)\right]$$

$$\frac{\gamma+1}{2\gamma}\frac{a^{*2}}{u_1}+u_1\left(\frac{\gamma+1}{2\gamma}-\xi\right)=\frac{\gamma+1}{2\gamma}\frac{a^{*2}}{u_2}+u_2\frac{\gamma+1}{2\gamma}$$

Multiply by $\frac{2\gamma}{\gamma+1}u_2$

$$\frac{a^{*2}}{u_2}u_2+u_2u_1\left[1-\frac{2\gamma\xi}{\gamma+1}\right]=a^{*2}+u_2^2$$

Hence, the equation in u_2:

$$u_2^2-u_2\left[\frac{a^{*2}}{u_1}+u_1-\frac{2u_1\gamma\xi}{\gamma+1}\right]+a^{*2}=0$$

This equation has the following form:

$$au_2^2+bu_2+c=0$$

$$a=1$$

$$b = -\left[\frac{a^{*2}}{u_1} + u_1 - \frac{2u_1\gamma\xi}{\gamma+1}\right]$$

$$c = a^{*2}$$

$$b^2 = \frac{a^{*4}}{u_1^2} + u_1^2 + \frac{4u_1^2\gamma^2\xi^2}{(\gamma+1)^2} + 2a^{*2} - \frac{4a^{*2}\gamma\xi}{\gamma+1} - \frac{4\gamma\xi\, u_1^2}{\gamma+1}$$

$$\Delta = b^2 - 4ac$$

$$\Delta = \frac{a^{*4}}{u_1^2} + u_1^2 - 2a^{*2} + \frac{4\,\gamma\xi}{\gamma+1}\left[\frac{u_1^2\gamma\xi}{\gamma+1} - a^{*2} - u_1^2\right]$$

$$\Delta = \frac{1}{u_1^2}\left(a^{*4} + u_1^4 - 2a^{*2}u_1^2\right) + \frac{4u_1^2\gamma\xi}{\gamma+1}\left[\frac{\gamma\xi}{\gamma+1} - \frac{a^{*2}}{u_1^2} - 1\right]$$

$$\Delta = \frac{1}{u_1^2}\left(a^{*2} - u_1^2\right)^2 + \frac{4u_1^2\gamma\xi}{\gamma+1}\left(\frac{\gamma\xi}{\gamma+1} - \frac{a^{*2}}{u_1^2} - 1\right)$$

$$u_2 = \frac{1}{2}\left[\frac{a^{*2}}{u_1} + u_1 - \frac{2u_1\gamma\xi}{\gamma+1} \pm \sqrt{\Delta}\right]$$

– the + sign corresponds to the Prandtl relation of the shock wave;

– the – sign corresponds to the dissipation of energy.

Next, the continuity equation gives:

$$\rho_2 = \frac{\rho_1 u_1}{u_2}$$

And the impulse equation is written as:

$$P_2 = P_1 + \rho_1 u_1^2(1-\xi) - \rho_1 u_2^2$$

NOTE.–

There is no simple relationship independent of the gas between coefficient ξ of the impulse equation and loss p used in definition of C_V of control valves.

7.2. Theoretical study of control valves, safety valves and gas pipelines

7.2.1. *Modeling control valves*

The model adopted includes the following elements as standard:

– a convergent;

– a singular loss of impulse;

– a divergent.

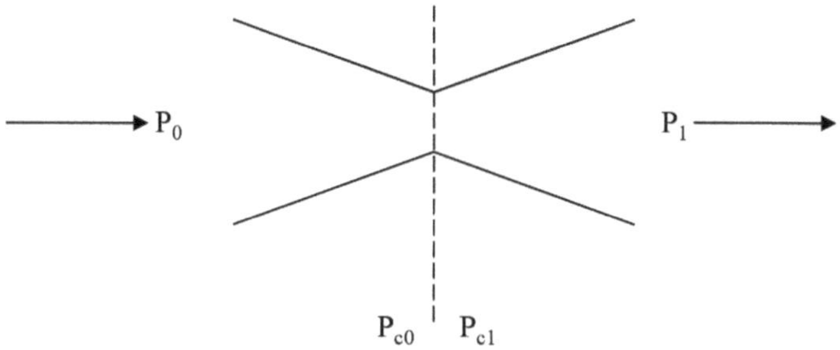

Figure 7.1. *Control valve model*

Let us suppose that the flow regime is both turbulent and subsonic everywhere. There is no friction in the convergent and, consequently, no dissipation of energy as heat. On real valves, this dissipation occurs only on exiting the minimum cross-section, that is, on entry to the divergent.

For an incompressible liquid, the pressure drop when the flow cross-section abruptly goes from A_0 to A_1 is as follows:

$$\Delta P = \xi \frac{\rho\, u_c^2}{2} \quad \text{with} \quad \xi = \left[1 - \frac{A_0}{A_1}\right]^2 \qquad [7.18]$$

For a gas, impulse is reduced by the amount:

$$\zeta \rho_c u_c^2$$

ρ_c and u_c are the density and velocity of the gas at the minimum cross-section of the path (at the *neck*) of the valve. There is no simple relationship independent of the gas between coefficients ξ and ζ, and nevertheless this equation [7.18] is traditionally employed not only for the calculation of control valves for liquids, but also for gases. This process, while incorrect in principle, leads certain manufacturers to indicate slightly different values for the valve coefficient C_V for gases and liquids. In any case, the industrial practice of using the same C_V for liquids and gases yields satisfactory results, on the condition, however, that an expansion coefficient is introduced for extreme expansions.

Finally, in the model proposed here, energy dissipation is localized in the plane perpendicular to flow and located at the minimum cross-section (the *neck*). We characterize the properties of the gas before and after crossing this plane with the indices c0 and cl.

We suppose here that there is no friction in the divergent. As in the convergent, the flow is isentropic.

EXAMPLE 7.1.–

$W = 0.07854$	$Y_0 = 0\ 0015374$
$D_0 = 0.1$ m	$P_0 = 0.554$ bar abs.
$A_0 = 0.007854$ m^2	$\rho_0 = 1$
$\gamma = 1.13$	$u_0 = 10$
$A_c = 0.0008$ m^2	$T_0 = 293$ K

In these data, we observe that cross-section A_c of minimum area is approximately 1/10 of the cross-section area A_0 in the pipe.

We have accepted a value of 0.6 for the impulse loss coefficient ζ.

Using these data, the conditions after the converger, that is, on entry into the minimum cross-section, are as follows:

$Y_{c0} = 0.1874326$

$P_{c0} = 0.4991$ bar abs.

$u_{c0} = 107.673474$

$\rho_{c0} = 0.911784456$

$T_{c0} = 289.5$ K

The calculation was performed using the isentropic flow equations with the variable cross-section.

On crossing the singular loss of impulse, the flow characteristics are as follows:

$Y_{c1} = 0.269819$

$P_{c1} = 0.414886364$ bar abs.

$u_{c1} = 128.8485$

$\rho_{c1} = 0.7619413$

$T_{c1} = 288.17$ K

After the isentropic crossing of the divergent:

$Y_1 = 0.00211238$

$P_1 = 0.4819966$ bar abs.

$u_1 = 11.4994$ m/s

$\rho_1 = 0.86961$ kg/m^3

$T_1 = 293.33$ K

By difference with the initial data:

– pressure drop:

$$P_0 - P_1 = 0.554 - 0.48199 = 0.072 \text{ bar}$$

– heating:

$$T_1 - T_0 = 293.33 - 293 = 0.33 \text{ K}$$

This slow expansion across a control valve is practically isothermal.

7.2.2. *Modeling valves*

A safety valve for subsonic flow is treated like a control valve. In general, however, at the minimum cross-section, the flow (at the *plug* or at the *neck*, as the reader chooses) is sonic, which interests us here.

If a valve is used to protect a tank at pressure P_0 and the valve opens to a space at pressure P_a (*a* for atmospheric), the overall expansion ratio is:

$$\tau = P_0/P_a$$

When the expansion rate reaches the critical value τ_c, the velocity at the plug is sonic and the flow no longer depends on the $P_0 - P_a$ difference, but only on the upstream conditions, and pressure P_0 in particular. Immediately after the plug, the regime is supersonic.

Consequently, a shockwave occurs at a specific (fixed) location between the valve and the exhaust pipe exit, *Upstream from this shockwave, flow is supersonic, while downstream it is subsonic.* Of course, the *closer* this wave is to the plug, the *higher* the pressure of the gas on arrival into the free space. In other words, the *closer* this wave is to the plug, the *lower* the difference between pressure P_0 of the protected device and pressure P_a of the free space receiving the exhaust.

When τ remains below a certain threshold, the $P_0 - P_a$ difference is moderated and the wave is localized to the valve divergent. This situation is more common, as demonstrated in the example below.

If τ increases, the wave moves downstream. In order to learn its position, we must proceed by iterations, performing various calculations. The position of the shockwave must be such that the gas pressure goes into the free volume equal to P_a.

When τ increases enough for the shockwave to reach the end of the exhaust pipe, the flow is supersonic from the valve to the exit. The calculation shows that the Mach number before the wave decreases and approaches 1 if τ increases or if we move downstream.

We know how to calculate the distance L^* from the valve, so that Y = 1.

If the length of the exhaust pipe is less than L^*, the flow is supersonic up to the shock wave and, if the pressure after the shockwave remains greater than P_a, while the shock wave is localized to the end of the pipe, the arrival of free air is accompanied by a noise due to the dissipation of the kinetic energy of the gas jet.

The situation in which the exhaust pipe length is greater than L^* does not physically exist. This means that the flow used for the calculation is excessive. If the flow is found to be reduced without changing the valve, the exhaust will be subsonic at the plug since the exhaust pipe resistance will have been sufficient to make the pressure increase on valve exit, so that the expansion ratio through the plug will be less than τ_c. Flow remains subsonic along the pipe until the exit. The acceptable flow will be weaker, as the pipe used will be longer.

Rather than accepting reduced flow, we can increase the diameter of the exhaust pipe. Indeed, this modification allows for an increase in the value of L^* for two reasons:

– the safety valve divergent is extended toward a greater section and the distance between Y and 1 increases, which in supersonic regime increases the friction parameter $\Delta X = X_1 - X_0$ corresponding to Y = 1 and Y_0 on valve exit.

– Length L^* is proportional not only to ΔX but also to pipe diameter D. Accordingly, it will strongly increase.

EXAMPLE 7.2.–

$P_0 = 17.5.10^5$ Pa $\quad$ $\gamma = 1.13$ $\quad$ $A_c = 0.002322\ \text{m}^2$

$T_0 = 303$ K $\quad$ $M = 44\ \text{kg.kmol}^{-1}$ $\quad$ $P_a = 1.19.10^5$ Pa

$L = 80$ m

The flow through the valve is certainly sonic because the critical expansion ratio is in the order of 2. For flow such as this, the classic valve calculation provides flow as mass:

$$W = 10.467542\ \text{kg.s}^{-1}$$

$$T_c = 203\times\frac{2}{2.13}=284.5\ \text{K}$$

$$P_c = 17.5.10^5\left(\frac{2}{2.13}\right)^{\frac{1.13}{0.13}}=9.8337.10^5\ \text{Pa}$$

$$\rho_c = \frac{44\times98337}{8314\times284.5}=18.2927\ \text{kg.m}^{t3}$$

$$u_c = \sqrt{\frac{1.13\times98337}{18.2927}}=246.467\ \text{m.s}^{t1}$$

$$Y_c = 1$$

After some trial and error, we can position the shockwave in the divergent part of the valve where the cross-section is $A_I = 0.05\ \text{m}^2$. The study of isentropic flow between cross-sections A_c and A_I provides the following results:

$Y_{I1} = 12.406952$

$P_{I1} = 9954.8354$ Pa

$u_{I1} = 666.56215\ \text{m.s}^{-1}$

$\rho_{I1} = 0.3141057 \text{ kg.m}^{-3}$

$T_{I1} = 167.7 \text{ K}$

After crossing the shockwave, conditions are as follows:

$Y_{I2} = 0.1294531$

$P_{I2} = 130433.5226 \text{ Pa}$

$u_{I2} = 91.1304815 \text{ m.s}^{-1}$

$\rho_{I2} = 2.2974856 \text{ kg.m}^{-3}$

$T_{I2} = 300.5 \text{ K}$

The exhaust pipe cross-section is calculated so that the gas flows through it at a velocity of 50 m.s^{-1} when taken at 0°C and 1 bar. Its density is then:

$$\rho_N = \frac{44}{22.72} = 1.9366197 \text{ kg.m}^{-3}$$

$$A_1^{(0)} = \frac{10.467542}{1.9366197 \times 50} = 0.1081 \text{ m}^2$$

$$DN = \sqrt{\frac{0.1081}{\pi / 4}} = 0.37109 \rightarrow DN350$$

$$A_1 = \frac{\pi}{4} \times 0.35^2 = 0.09616 \text{ m}^2$$

The isentropic study between cross-sections A_I and A_1 provides the following results:

$Y_1 = 0.0315512$

$P_1 = 137812.0012 \text{ Pa}$

$u_1 = 45.13256669 \text{ m.s}^{-1}$

$\rho_1 = 2.412137054 \text{ kg.m}^{-3}$

$T_1 = 302.3623 \text{ K}$

The calculation of the cutoff length gives:

$L^* = 279 \text{ m}$

The distance separating the valve from the flare drum is taken as equal to 80 m (so, less than L^*).

Hence:

$$\Delta X = \frac{0.03 \times 80}{2 \times 0.35} = 3.4285$$

For this value:

$Y_2 = 0.042311$

$P_2 = 118964 \text{ Pa}$

$u_2 = 52.24 \text{ m.s}^{-1}$

$\rho_2 = 2.083 \text{ kg.m}^{-3}$

$T_2 = 302.25 \text{ K}$

7.2.3. *Conclusions*

– The required pressure of 1.19 bars is very close to the pressure observed (1.189 bar).

– The gas, which had a temperature of 303 K, is very close to the temperature observed (302.25 K).

– The variation in gas temperature between exhaust pipe entry and exit is completely negligible (302.36 – 302.25 = 0.11 K). In practice, we perform the calculation of expansion by friction in a pipe by accepting that the temperature remains constant. Valve exhaust is an example of this.

– On the other hand, the plug temperature is very low (167.7 K = –105.45°C), and hence, there is a possibility of icing.

7.2.4. *Pressure drop in a gas pipeline*

The expressions that have been established for the expansion of gas in a pipeline with friction are quite suitable for expressing the behavior of a gas in a long conduit, such as a gas pipeline.

EXAMPLE 7.3.–

Here, we consider natural gas as pure methane, so, at an initial pressure of 35 bar abs. and a temperature of 60°C, the properties will be:

$$\rho = \frac{16 \times 35.10^5}{8314 \times 333} = 20.22 \text{ kg.m}^{-3}$$

$$\mu = 11.10^{-6} \text{ Pa.s} \qquad \gamma = 1.31$$

Pipeline diameter is 0.6 m, transporting 300,000 $\text{Nm}^3.\text{h}^{-1}$

$$W = \frac{300\,000 \times 16}{22.42 \times 3600} = 59.74 \text{ kg.s}^{-1}$$

Entry velocity is:

$$V_0 = \frac{59.74}{20.22 \times (\pi / 4) \times (0.6)^2} = 10.4 \text{ m.s}^{-1}$$

The square of the initial Mach number is:

$$Y_0 = \frac{20.22 \times 10.4^2}{1.31 \times 35.10^5} = 4.76.10^{-4}$$

With the accepted friction coefficient being equal to 0.03 for a length of 20 km, we obtain:

$$X_1 - X_0 = \frac{0.03 \times 20000}{2 \times 0.6} = 500$$

Hence, the equation:

$$\frac{1}{Y_1} = -2\times 1.31\times 500 + \frac{1}{4.76.10^{-4}} + \text{Ln}\left[\frac{1+\dfrac{2}{0.31Y_1}}{1+\dfrac{2}{0.31\times 4.76.10^{-4}}}\right]^{1.151}$$

$Y_1^{(0)}$=0.0012644

$Y_1^{(1)}$=0.00126

$Y_1^{(2)}$=0.0012662

$Y_1^{(3)}$=0.00126628

$$P_1 = 35.10^5\left[\frac{4.76.10^{-4}\left(4.76.10^{-4}+\dfrac{2}{0.31}\right)}{12.6628.10^{-4}\left(12.6628.10^{-4}+\dfrac{2}{0.31}\right)}\right]^{1/2}$$

$$P_1 = 21.4575 \text{ bar}$$

$$u_1 = 10.4\left[\frac{12.6628.10^{-4}\left(4.76.10^{-4}+\dfrac{2}{0.31}\right)}{4.76.10^{-4}\left(12.6628.10^{-4}+\dfrac{2}{0.31}\right)}\right]^{0.5}$$

$$u_1 = 16.962 \text{ m.s}^{-1}$$

$$\rho_1 = 20.22 + \frac{10.4}{16.962} = 12.397 \text{ kg.m}^{-3}$$

$$T_1 = \frac{16\times 2145700}{8314\times 12.397} = 333 \text{ K} = 60°\text{C}$$

As expansion is isothermal, it should suffice to use the simplest method indicated for exhaust pipes.

NOTE.–

The expanded gas must be periodically recompressed every 30–50 km. This heats the gas, which is subsequently cooled naturally in the ground, since gas pipelines are underground. The overall thermal transfer coefficient is in the order of 10 W/m^2°C for heat exchange with the ground. This coefficient is probably increased if the ground is wet.

If T_0 and T_1 are the temperatures of the gas at the beginning and at the end of a segment, we can assess the isothermal pressure drop at the average temperature:

$$T_m = \frac{1}{3}(2T_0 + T_1)$$

7.2.5. *Overall conclusions*

We know that the energy equation is:

$$H + C = \text{const.}$$

This equation is precise, irrespective of friction.

The gas velocity varies slightly or not at all:

– between upstream and downstream of a slow expansion control valve;

– between a device protected by a safety valve and the atmosphere;

– even in a pipeline in which gas expands from 35 to 21.5 bar, and where velocity varies from 10.4 to 17 m.s^{-1}, the temperature remains almost constant.

Thus, the main source of temperature variation in these three cases remains the difference separating the gas from an ideal state, which justifies the interest of the Joule–Thomson expansion study.

8

Safety Valves and Rupture Disks

8.1. Pressure around a safety valve

8.1.1. *Operating pressure of a protected device*

Pressure P_S is used to determine the dimensions of this device.

8.1.2. *Maximum operating pressure*

Accounting for random events susceptible to a rise during operation, P_{MS} is the pressure to which the device may be subjected for a short time.

The assessment of this pressure *must be done on a case-by-case basis*. Accordingly, a stripping column operating at $P_S = 2.5$ bar abs. and supplied with steam at 6 bar abs. through a control valve would have a P_{MS} equal to 6 bar abs., which corresponds to the valve fixed wide open (here, we disregard the pressure drop through the valve).

Here, we see the danger of arbitrary rules such as those that would indicate that:

$$P_{MS} = 1.1\ P_S \qquad (\text{in barG})$$

8.1.3. *Pressure calculation*

This is pressure P_C used for wall thickness calculations, for selecting the welding procedure, etc., and in general, for everything concerning the mechanical resistance of devices.

The calculation codes of various countries define the way in which we obtain pressure P_C from the maximum operating pressure P_{MS}. Thus, it is a widespread practice to use:

$$P_C = 1.1\, P_{MS} + 1.4 \qquad (\text{barG})$$

This expression applies for $1 < P_{MS} < 20$ barG.

8.1.4. *Set pressure*

At pressure P_T the valve lifts from its seat slightly (*rises* from its seat) and lets a small flow of fluid pass through. As the pressure increases in the device, the valve lifts more and more, that is, it lifts higher, and simultaneously increasing the evacuated flow.

Pressure P_T compensates exactly:

– The force of the spring holding the plug on its seat. This force is a result of the preset calibration of the spring's compression. However, we will see that the situation is more or less complex depending on the type of the valve used.

– The weight of the moving parts, since these valves are always mounted vertically. If the plug is simply weighted, the weight of a given mass replaces the spring and the valve is kept closed merely by the force of gravity.

For a gas, we typically choose a value of P_T equal to the device's pressure calculation P_C rather than P_{MS}. Indeed, it is worth maximizing P_T, as this reduces the size of the valve and the connecting pipes, helping to reduce costs. Moreover, this helps the valve avoid the need to open too frequently; indeed, too frequent opening ultimately damages the seal of the plug on its seat, compromising the equipment's overall reliability while valve exhaust also creates unnecessary noise pollution for the environment.

For liquids, the majority of valves installed are thermal expansion valves and, as we will see, for P_T, we must choose the value that corresponds to P_{MS}. This is not the case for the discharge of a volumetric pump. The pump manufacturer should provide the maximum pressure at which the machine can operate, and it is quite possible to choose an intermediate value between this pressure and the pressure of the calculation resulting from the characteristics of the discharge line as set pressure. Indeed, the difference between these two pressures is often significant.

8.1.5. *Pressure upstream of the safety valve*

It is pressure P_{UP} that governs upstream of the valve, not when the plug is lifted, by when the valve frees the flow for which it was calculated, that is when the plug is totally lifted. Then, pressure P_{UP} is given by:

$$P_{UP} = P_T + \Delta P_A$$

ΔP_A is the over-pressure accumulated due to the friction incurred by the fluid to maintain the valve completely open.

We define an accumulation rate α, which is:

$$\alpha = \Delta P_A / P_T$$

– For a single valve on gas, we must, according to the norm API 520, have:

$$\alpha \leq 0.1$$

– For several valves on a gas, we must, according to the same norm, use:

$$\alpha \leq 0.16$$

Therefore, Sapag valves used on gases are such that:

$$\alpha = 0.05$$

If $P_T = P_C$, this accumulation can be tolerated easily by the device.

On the other hand, for a liquid, we have:

$$\alpha = 0.25$$

We understand then that it is more wise to use:

$$P_T = P_{MS} < P_C$$

With:

$$P_C \geq 1.25\ P_{MS}$$

8.1.6. *Closing differential (drop)*

This is difference ΔP_F between the set pressure and the value of the upstream pressure (value below the set pressure) for which the valve closes.

Most valves have an adjustment mechanism for valve closure calibration.

Often:

$$\Delta P_F \# 0.05\ P_T$$

8.1.7. *Counter-pressure downstream of the valve*

This is the static pressure P_{DS} immediately downstream of the valve.

We can observe:

– the *pre-existing* (or permanent) counter-pressure before valve opening;

– the counter-pressure *created after* valve opening. This is higher than the previous pressure due to friction in the exhaust pipe, and can also increase if the exhaust occurs in an intermediate capacity where the freed gas accumulates temporarily.

NOTE.–

The boiler pressure defined by French norms can be found by reading the plate affixed to the machine. This should read:

$$P_{MS} < \text{Boiler pressure} < P_C$$

8.2. Choice between two types of safety valves

8.2.1. *Usual valves*

In choosing a valve, the aim is to balance the set pressure with the force of the spring and/or the weight of the moving parts. If F_T is the sum of these forces:

$$F_T = P_T$$

The usual type of valve corresponds to the most straightforward and cost-effective design as represented in Figure 8.1.

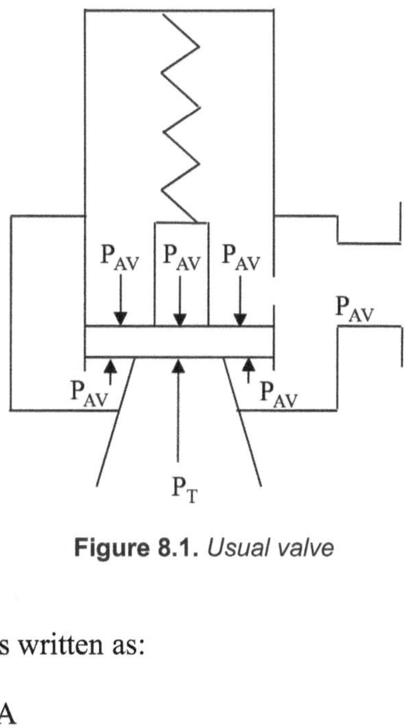

Figure 8.1. *Usual valve*

The plug balance is written as:

$$P_T A = F_T + P_{AV} A$$

Upstream pressure P_T (set pressure) is not independent of counter-pressure P_A and increases with it. Hence, the uncertainty surrounding P_T once we have adjusted the spring and set the value of F_T. The so-called *balanced* plug was developed in order to alleviate this inconvenience.

8.2.2. *Balanced plugs*

The principle is described in Figure 8.2.

The plug is subjected to two pressures:

– the lower side withstands pressure P_T spread over surface A;

– the upper side withstands atmospheric pressure spread over the same surface A. In other words, the cross-section of the passage to the right of the plug is equal to A. A vent opens the bellows to the atmosphere via the interior of the valve cap.

In absolute pressure, the plug balance is written as:

$$P_T A = F_T + P_{ATM} A$$

As atmospheric pressure P_{ATM} changes only to a very limited extent over time, F_T depends only on P_T.

The sides of the bellows are subjected to atmospheric pressure (inner side) and the downstream counter-pressure (outer side). The difference between these forces must be compatible with the mechanical resistance of the bellows.

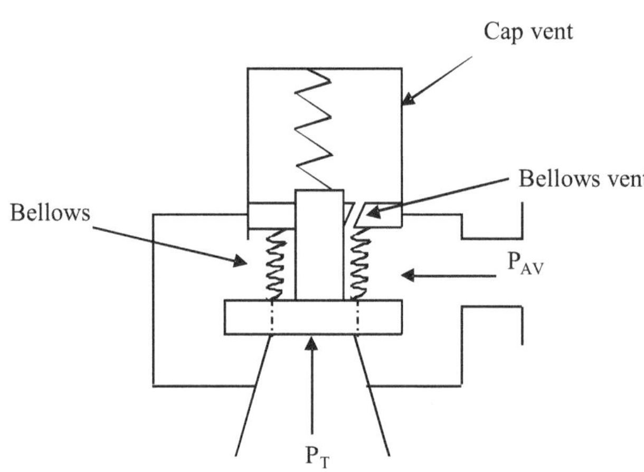

Figure 8.2. *Balanced valve*

8.2.3. *Choosing a safety valve type*

According to what we have seen above, we know that:

1) valves going into the atmosphere are always of the conventional (usual) type;

2) valves going into a process space are always of the balanced type. As we have seen, the mechanical resistance of the bellows is then dependent on the downstream counter-pressure and we consequently indicate the value of this pressure in all cases, even if the fluid flow is critical.

8.2.4. *Inlet cross-section norms*

Typically, the following range is used (Table 8.1).

Type	D	E	F	G	H	J	K	L
Cross-section (cm^2)	0.710	1.264	1.980	3.244	5.068	8.301	11.85	18.40
Type	M	N	P	O	R	T	V	W
Cross-section (cm^2)	23.22	27.99	41.15	71.27	103.2	167.7	258.3	387.2

Table 8.1. *Inlet norms*

NOTE.–

The accumulation and the drop distance in piloted valves is close to zero, resulting in a considerably greater precision in these valves. However, we will not describe the various mechanisms used for this category of valves here.

8.3. Relationship between flowrate and pressure

8.3.1. *Gases*

The mass flowrate of a gas through the valve inlet is:

$$W = A\,U_C\,\frac{M}{V_C}$$

A: inlet cross-section (m^2)

U_C: gas velocity to the right of the inlet ($m.s^{-1}$)

M: molar mass of gas ($kg.kmol^{-1}$)

V_C: molar volume of gas to the right of the inlet ($m^3.kmol^{-1}$)

Let us multiply the second expression of this equation by 1:

$$1 = \frac{P_1 V_1}{Z\,R\,T_1}$$

Index 1 characterizes the entry conditions of the valve and Z is the compressibility factor.

Hence:

$$W = A\frac{V_1}{V_C}U_C\frac{P_1 M}{Z\,R\,T_1}$$

But, for adiabatic expansion:

$$\frac{V_1}{V_C} = \left(\frac{P_C}{P_1}\right)^{1/\gamma} = r^{1/\gamma}$$

where r is the expansion ratio.

Hence:

$$W = Ar^{1/\gamma}\sqrt{\frac{M}{Z\,R\,T_1}}U_C P_1\sqrt{\frac{M}{Z\,R\,T_1}}$$

We can disregard the square of velocity on valve entry (this velocity is in the order of 10 $m.s^{-1}$) in favor of the square of velocity to the right of the inlet, as the latter velocity is around ten times higher. We can then obtain an expression of U_C^2 according to r:

$$\frac{\rho_1}{P_1}\times U_C^2 = \frac{2\gamma}{\gamma-1}\left[1-r^{\frac{\gamma-1}{\gamma}}\right]$$

We find justification for this expression in the work of Brun *et al.* [BRU 68, p. 360]. If we obtain this result, we can use the following expressions:

– conservation of matter;

– impulse equation;

– Laplace equation;

– ideal gas state equation.

On the other hand:

$$\frac{M}{Z\,R\,T_1} = \frac{M}{P_1 V_1} = \frac{\rho_1}{P_1}$$

Finally:

$$W = A\, r^{1/\gamma} \sqrt{\frac{2\gamma}{\gamma - 1}\left(1 - r^{\frac{\gamma-1}{\gamma}}\right)} P_1 \sqrt{\frac{M}{Z\,R\,T_1}}$$

Now, let us suppose that the expansion is critical, that is, where the gas flow is sonic at the valve opening. The expansion rate is then:

$$r = \left(\frac{2}{\gamma + 1}\right)^{\frac{\gamma}{\gamma-1}}$$

(see Brun *et al.* [BRU 68, p. 389]).

And, replacing r by this expression, we have:

$$W = A\sqrt{\gamma\left(\frac{2}{\gamma+1}\right)^{\frac{\gamma+1}{\gamma-1}}}\; P_1\sqrt{\frac{M}{Z\,R\,T_1}}$$

EXAMPLE 8.1.–

$P_1 = 10.34$ bar abs $= 1.034.10^6$ Pa $\quad M = 58$ kg.kmol^{-1}

$A = 8.3.10^{-4}$ m^2 $\quad \gamma = 1.3$

$T_1 = 353.8$ K $\quad Z = 0.79$

$$W = 8.3.10^{-4}\sqrt{1.3\left(\frac{2}{2.3}\right)^{\frac{2.3}{0.3}}}\ 1.034.10^6\sqrt{\frac{58}{0.79 \times 8314 \times 353.8}}$$

$$W = 2.87 \text{ kg.s}^{-1}$$

8.3.2. *Liquids*

The basic expression for flow of a liquid through an opening is:

$$W = 0.62\ A\sqrt{2\rho_L \Delta P} \qquad [8.1]$$

W: liquid flowrate in mass (kg.s^{-1})

A: liquid cross-section in mass (m^2)

ρ_L: liquid density (kg.m^{-3})

ΔP: pressure drop on crossing inlet (Pa)

The dimensionless coefficient 0.62 announces the contraction of the process flow immediately after crossing the opening.

For an upstream pressure equal to the set pressure, the liquid only begins to flow at a rate close to zero. The valves are oversized for an accumulation ratio equal to 25% of the set pressure P_T so that we must write:

$$\Delta P = 1.25 P_T - P_{DS}$$

P_{AV} is the counter-pressure downstream of the valve.

In this last expression, the pressures are expressed relative to the atmosphere. In addition, this expression of ΔP is only applicable for a conventional valve whose inlet cross-section needs to be determined by means of equation [8.1].

The manufacturers affirm that the cross-section of a balanced valve must be increased by division of a coefficient less than 1, when the downstream counter-pressure exceeds the set pressure by a value in the order of 15%. Thus, the counter-pressure influence for such valves is greater than it would be for a conventional-type valve. Consequently, balanced valves are of limited interest here.

EXAMPLE 8.2.–

Therefore, in order to calculate the cross-section of a conventional valve on a liquid:

$$P_T = 12 \text{ barG} = 12.10^5 \text{ Pa} \qquad P_{AV} = 0 \text{ barG}$$

$$Q = 5.10^{-4} \text{ m}^3.\text{s}^{-1} \qquad P_L = 996 \text{ kg.m}^{-3}$$

$$A = \frac{5.10^{-4}\sqrt{996}}{0.62\sqrt{2 \times 1.25 \times 12.10^5}} = 0.147.10^{-4} \text{ m}^2$$

We would choose a D-type valve ($A = 0.71$ cm^2, see Table 8.1).

8.3.3. *Viscous liquids*

The transportation of these liquids is always performed by volumetric discharge pumps on which a valve is fitted.

The designated flow rate through the valve is that of the pump if it is a constant flow pump. This is equal to the maximum flow if the machine is of periodical flow (piston pump).

The influence of viscosity is applied by means of the Reynolds number, which is defined as:

$$Re = \frac{4W}{\mu P}$$

W: mas flowrate ($kg.s^{-1}$)

μ: viscosity (Pa.s)

P: inlet perimeter (m)

For a circular inlet of cross-section A:

$$P = \pi D = \pi\sqrt{\frac{4A}{\pi}} = \sqrt{4\pi A}$$

Hence:

$$Re = \frac{4W}{\sqrt{4\pi}\mu\sqrt{A}} = 1.128\frac{W}{\mu\sqrt{A}}$$

The expression of the valve cross-section on a liquid is (in coherent units):

$$A = \frac{W}{0.62\ K_v\sqrt{2\rho_L\left(1.25P_T - P_{AV}\right)}}$$

According to norm API 520, the K_v coefficient is an increasing function of the Reynolds number. Among the equations in [8.1], we find an approximate expression of the $K_v = f\,(Re)$ curve of norm API 520.

$$\left.\begin{array}{lll} 10 < Re \leq 250 & K_v = & 0{,}292\ LnRe - 0{,}746 \\ 250 < Re \leq 10000 & K_v = & 0{,}051\ LnRe + 0{,}524 \\ Re > 10000 & K_v = & 1 \end{array}\right\} \qquad (21.1)$$

NOTE.–

1) For gases, the accumulation ratio is clearly lower than that for liquids; however, we have to use:

$$P_1 = 1.05\ P_T$$

2) We have seen that the valve calculation must be performed while accounting for a certain accumulation ratio, that is, in a way such that the plug rises completely. If $\varnothing_o$ is the inlet diameter, the lateral cross-section open to the fluid path is, for a plug rise h:

$$\varnothing_o \, h \sin\theta$$

This cross-section is called the curtain cross-section.

θ: incline angle of the seat range to vertical.

And, for a complete plug lift, the curtain cross-section must be equal to the inlet cross-section:

$$A = \frac{\pi}{4}\varnothing_o^2 = \pi \, \varnothing_o \, h \sin\theta$$

Hence:

$$h = \frac{\varnothing_o}{4 \sin\theta}$$

8.3.4. *Flow through valves*

We will deal with the following examples here:

1) Semi-enclosed space (open entry, closed exit):

– Holding tank: The exhaust flow from the gaseous sky is equal to the flow of the supplied liquid.

– Condenser: the condensate is not evacuated. Consequently, the device fills with liquid and the pressure in the vapor space increases until reaching the set P_T. The exhaust flow will then be identical to that of the vapor produced. This also applies if the non-condensed gases are not evacuated, as condensation occurs with increasing difficulty. We should recall that a condenser can manage the vapors of a gas–liquid column or those of an evaporator. Finally, any heating system using steam is a particular condenser.

2) The circulation of cold fluid in an exchange is interrupted:

– The cold liquid tends to take its saturation temperature at a pressure equal to the set pressure P_T. From the moment when the valve escapes, we can write *the classic equations* of a heat exchanger:

$$Q = US\Delta T = WC\left(T_e - T_s\right) \qquad \left(\Delta T = LMTD\right)$$

$$t_e = t_s = t\left(P_T\right)$$

This system of two equations allows for the calculation of temperature t of the stagnant cold fluid and exit temperature T_s of the hot fluid. The thermal transfer is then:

$$Q = US\Delta T = Lw$$

L is the vaporization heat of the cold fluid from which we deduce the flowrate w of vapor escaping to the valve.

– A reactor with an exothermal reaction is no more cooled. The reaction can go out of control, but this requires a precise study of the process used.

3) Rupture of an exchanger pipe:

If ΔP is the difference between operating pressure of the two fluids and if A is the contact area of the two fluids, we obtain the leak flow by applying the safety valve formulae to high-pressure fluids. This leak flow, expressed in volume, allows for the valve calculation of a low-pressure fluid. The set pressure of this valve will of course be less than or equal to the calculated pressure on the low-pressure side of the exchanger.

4) Control valve failure:

We take the flow of the fully open valve if it is a supply valve. On the other hand, if it is an outlet valve, we will consider the consequences of its closure.

5) Incorrect operation:

We will take the most extreme conditions of flow and pressure.

8.4. Upstream and downstream connections

8.4.1. *Connection between valve and protected device*

The pressure drop in this line must not exceed 3% of the set pressure (expressed in gauge bars). Indeed, a higher value would remove pressure from the set pressure. Furthermore, the vibrations created by the intense turbulence are always damaging to a structure's integrity.

On discharge from a volumetric piston machine, the pressure fluctuations can provoke, with a fast rhythm, valve opening and closing (which can quickly alter its reliability) if the two following conditions are not met:

– maximizing the distance between the valve and the protected machine;

– maximizing the difference between the operating pressure of the line and the set pressure of the valve.

8.4.2. *Exhaust pipe*

The set pressure is imposed together with pressure P_l at the exit of the exhaust line (often the atmosphere). Counter-pressure P_{DS} at the valve exit flange must be such that the flowrate is the same through both the valve and the line. Hence, we have two equations:

– that of the valve (that we know);

– that of the exhaust line (that we must establish).

Thus, we can calculate counter-pressure P_{DS} and the exhaust flowrate, so that the notion of *critical flowrate* has no meaning here.

Exhaust pipe diameter must be equal or greater than the valve exit flange. Consequently, it is necessary for the flow to occur as freely as possible, and in practice, we must limit the exit pressure drop ΔP_{DS} as follows:

– Conventional valves:

$\Delta P_{DS} < 0.1\ P_T$ (operation hazards)

$\Delta P_{DS} < 0.2\ P_T$ (fire hazards)

– Balanced valves:

$\Delta P_{AV} < 0.5\ P_T$

The exit pressure drop is not only due to pipe friction, but also due to the acquisition of kinetic energy by the expanding gas.

The friction term on length dL is:

$$dP_F = -f\frac{\rho V^2}{2}\frac{dL}{D} = -f\frac{G^2}{2\rho}\frac{dL}{D}$$

If ρ_{DS} and P_{DS} are the density and gas pressure at the valve exit flange and ρ and P the density and gas pressure along the pipe, supposing isothermal expansion, we can write:

$$\rho = \rho_{DS}\frac{P}{P_{DS}}$$

This calculation effective shows that the pressure drop obtained is somewhat more significant if we use the hypothesis of an isothermal expansion than supposing adiabatic expansion. Hence:

$$dP_F = -f\frac{G^2}{2\rho_{DS}}\frac{P_{DS}}{P}\frac{dL}{D}$$

On the other hand, the variation in flow movement amount is (if W is the gas flowrate in mass):

$$W\ dV = W\ d\left(\frac{G}{\rho}\right) = W\ G\ d\left(\frac{1}{\rho}\right)$$

This force corresponds to the pressure variation:

$$dP_C = -\frac{W}{A}G\ d\left(\frac{1}{\rho}\right) = -G^2\ d\left(\frac{1}{\rho}\right) = -\frac{P_{DS}G^2}{\rho_{DS}}d\left(\frac{1}{P}\right)$$

Therefore, along pipe length L, the total pressure variation is:

$$dP = dP_F + dP_C = -\frac{P_{AV}G^2}{\rho_{AV}}\left[\frac{f}{2\,D}\frac{dL}{P} - \frac{dP}{P^2}\right]$$

Let us write out that:

$$\pi_{AV} = \frac{P_{AV}G^2}{\rho_{AV}} = \frac{P_1G^2}{\rho_1} = \pi_1 = \pi \qquad \left(Pa^2\right)$$

Index 1 characterizes the downstream end of the exhaust line. After integration, we obtain:

$$P_1^2 = P_{DS}^2 - \frac{\pi f L}{D} + 2\pi Ln\frac{P_1}{P_{DS}}$$

This equation allows for the calculation of P_{DS} according to P_1 and vice versa. This is resolved by successive approximations. For example:

$$\left(P_{DS}^{(0)}\right)^2 = P_1^2 + \frac{\pi\, fL}{D}$$

$$\left(P_{DS}^{(1)}\right)^2 = P_1^2 + \frac{\pi\, fL}{D} - 2\pi Ln\frac{P_1}{P_{DS}^{(0)}}$$

We also need to verify that the velocity attained by the expanded gas remains below the velocity of sound.

The gas density can be obtained by:

$$\rho_1 = \frac{M\,P_1}{R\,T}$$

The velocity of sound is (see the Appendix):

$$a = \sqrt{\gamma\frac{P_1}{\rho_1}}$$

Hence, the condition required for velocity V_1:

$$V_1 = \frac{G_1}{\rho_1} < \sqrt{\gamma \frac{P_1}{\rho_1}}$$

That is:

$$\rho_1 V_1^2 = \frac{G_1^2}{\rho_1} < \gamma P_1$$

If, for example, as we often see:

$$P_1 = 10^5 \text{ Pa, we must have: } \rho_1 V_1^2 < \gamma.10^5 \text{ Pa}$$

EXAMPLE 8.3.–

$P_1 = 10^5$ Pa $\qquad \gamma = 1.4$

$G = 155$ kg.m^{-2}.s^{-1} $\qquad M = 29$ kg.kmol^{-1}

$L = 20$ m $\qquad T = 293$ K

$D = 0.1$ m $\qquad f = 0.03$

$$\rho_1 = \frac{29 \times 10^5}{8314 \times 293} = 1.19 \text{ kg.m}^{-3}$$

$$\pi = \frac{155^2 \times 10^5}{1.19} = 2.02.10^9 \text{ Pa}^2$$

$$\left(P_{DS}^{(0)}\right)^2 = 10^{10} + 2.02.10^9 \times 0.03 \times \frac{20}{0.1} = 2.212.10^{10} \text{ Pa}^2$$

$$P_{DS}^{(0)} = 1.49.10^5 \text{ Pa}$$

$$\left(P_{DS}^{(1)}\right)^2 = 1.212.10^{10} + 2 \times 2.02.10^9 \text{ Ln} \frac{1.49}{1}$$

$$P_{DS}^{(1)} = 1.17.10^5 \text{ Pa}$$

A third iteration provides:

$$P_{AV}^{(2)} = 1.129.10^5\,\text{Pa}$$

We must check that:

$$\frac{G_1^2}{\rho_1} < \gamma P_1 \qquad \text{that is:} \qquad \frac{155^2}{1.19} < 1.4.10^5$$

$$20189 < 140000$$

8.4.3. *Reaction force on the structure*

This force is the result of the mass flow of the gas by the velocity reached. This is momentum flow.

In general, flow through the valve is critical and it is here that reaction is significant. We have seen that:

$$W = A \sqrt{\gamma \left(\frac{2}{\gamma + 1} \right)^{\frac{\gamma+1}{\gamma-1}}} \; P_1 \sqrt{\frac{M}{Z\,R\,T_1}}$$

For critical flow, the inlet temperature of the valve is:

$$T_0 = T_1 \left(\frac{2}{\gamma + 1} \right)$$

Hence, the speed of sound at the inlet, that is, the speed of the gas, is:

$$a_o = \sqrt{\frac{\gamma R\,T_o}{M}}$$

And the reaction force is written simply as:

$$F_R = W\,a_o$$

But this value must be increased to account for the vibration frequency of the structure itself. The result obtained should cover all of these effects if we add a multiplier coefficient equal to 2.

EXAMPLE 8.4.–

$$P_1 = 14.2.10^5 \text{ Pa} \qquad Z = 1$$

$$T_1 = 293 \text{ K} \qquad M = 29$$

$$A = 0.0168 \text{ m}^2 \text{ (type T)} \quad \gamma = 1.4$$

$$W = 0.0168\sqrt{1.4\left(\frac{2}{2.4}\right)^{\frac{2.4}{0.4}}} \times 14.2.10^5\sqrt{\frac{29}{8314\times 293}}$$

$$W = 56.4 \text{ kg.s}^{-1}$$

$$T_o = 293\times\frac{2}{2.4} = 244 \text{ K}$$

$$a_o = \sqrt{\frac{1.4\times 8314\times 244}{29}} = 312 \text{ m.s}^{-1}$$

$$F_R = 56.4\times 312\times 2 = 35194 \text{ N}$$

That is a 3.58 tonne-force through division by $g = 9.81 \text{ m.s}^{-2}$.

8.5. Various applications

8.5.1. *Vacuum-breaking valves*

The inlet diameter of these valves is large: from 50 to 350 mm, and the set negative pressure can be selected between 2.2 and 350 mbar.

The plug is held in closed position by a spring, but for low negative pressures, it can be simply weighted (see Figure 8.3).

Unlike in most pressure valves, flow through the open valve occurs in non-critical regime.

The use of vacuum-breaking valves is recommended in the following cases:

– stopping a reboiler or heating of a distillation column;

– stopping the vapor injection in a stripping column;

– stopping the heating of an evaporator;

– and, in general, when cooling a device that operates under a pressure greater than or equal to atmospheric pressure and containing a liquid whose vapor pressure at ambient temperature is less than the atmospheric pressure (this is, notably the case for water).

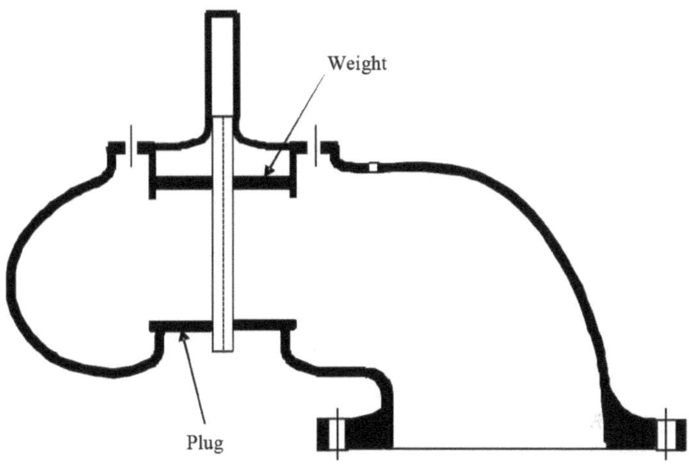

Figure 8.3. *Principle of a vacuum-breaking valve with weighted plug*

EXAMPLE 8.5.–

$P = 50$ mbar $\gamma = 1.4$ $M = 29$ kg.kmol^{-1}

$r = 0.95$ $P_1 = 1.013.10^5$ Pa $T = 293$ K

inlet diameter: 0.125 m

Flowrate of air in:

$$W = \frac{\pi}{4}(0.125)^2 \times 0.95^{\frac{1}{1.4}} \sqrt{\frac{2\times 1.4}{0.4}\left(1-0.95^{\frac{0.4}{1.4}}\right)} \times 1.013.10^5 \sqrt{\frac{29}{8314\times 293}}$$

$$W = 1.31\,m^3.s^{-1}$$

8.5.2. *Breathing of reservoirs under atmospheric pressure*

When storing a volatile liquid, we prevent it from evaporation by reducing its contact with the atmosphere. Accordingly, the reservoir is closed. However, it can be held in negative pressure in the following two cases:

– draining the liquid;

– reduction of the ambient temperature, that is, reduction of the vapor pressure of the liquid.

A reservoir placed under negative pressure can easily become deformed and thereby be rendered unusable.

Inversely, on filling or on increasing the exterior temperature, the reservoir can be placed under pressure.

In order to keep the vapor space above the liquid within the acceptable pressure range, manufacturers propose breathing valves. These valves open:

– for a negative pressure than can be set between 2 and 6.5 mbar;

– for an over-pressurization between 2.5 and 100 mbar.

The inlet diameters can be chosen between 50 and 300 mm. These valves include two distinct plugs: one for negative pressure and another for over-pressure.

8.5.3. *Liquid escaping without free surface (low vapor pressure)*

This situation occurs when the flow of a coolant is interrupted in a liquid-phase chemical reactor.

If the thermal power created is P and if Ω is the volume of confined liquid, the temperature increase over the time lapse $d\tau$ is:

$$dT = P\ d\tau/\Omega\rho_L C_L$$

ρ_L and C_L are the density and the thermal capacity of the liquid.

The creation of volume $d\Omega$ ensuing from this temperature increase is:

$$d\Omega = 3\lambda\Omega dt$$

λ is the linear dilatation coefficient and 3λ the cubic dilatation coefficient.

By eliminating dt between these two equations, we obtain:

$$\frac{d\Omega}{d\tau} = \frac{3\lambda P}{\rho_L C_L}$$

This flow must be evacuated by a thermal expansion valve.

In the examples provided here, we have disregarded the expansion of the metal surrounding the liquid.

EXAMPLE 8.6.–

$P = 1.6$ W $\qquad \rho_L = 950$ kg.m^{-3}

$3\lambda = 0.00133$ K^{-1} $\qquad C_L = 4\ 186$ J.kg^{-1}.K^{-1}

$$\frac{d\Omega}{d\tau} = \frac{0.00133 \times 1.6}{950 \times 4186} = 54\ \text{L .s}^{-1}$$

8.6. Rupture disks

8.6.1. *Rupture pressure*

On bursting, a rupture disk instantly acquires the maximum passage cross-section necessary for the fluid to escape, which occurs when the difference between the upstream pressure and the downstream pressure has a precise value known as the rupture pressure P_E.

In each batch produced, the manufacturer performs destructive tests at a given temperature (for example, four disks are subjected to the test). At this temperature, we obtain:

– the mean value $\overline{P}_E$ of P_E;

– tolerance around this average.

So that:

$$\overline{P}_E - \Delta P_{EI} < P_E < \overline{P}_E + \Delta P_{ES}$$

The lower and upper tolerances ΔP_{EI} and ΔP_{ES} are seldom less than $0.04.\overline{P}_E$. We often have:

$$\Delta P_{EI} = 0.04.\overline{P}_E \quad \text{and} \quad \Delta P_{ES} = 0.07\ \overline{P}_E$$

(0.04 and 0.07 are the relative upper and lower relation tolerance limits).

If the rupture pressure is very low, the tolerance limits can be considerably higher. As a general rule, these limits become less, as P_E increases.

Tests at different temperatures provide the (decreasing) P_E (T) function, which allows manufacturers to guarantee an average $\overline{P}_E$ value for a given operating temperature T_S.

For constant tolerance levels, the lower limit of P_E is minimum for the maximum value of the nominal disk diameter (50 mbar for DN 600, for instance). However, the upper limit of P_E is maximum if the DN is minimum (7,000 bars for DN 10 as non-standard).

8.6.2. *Operating ratio*

This is the ratio of operating pressure P_S to rupture pressure P_E. This P_S/P_E ratio below 1 must not exceed a certain level that varies according to the type of disk in order to avoid metal fatigue, which can harm the disk reliability (see Table 8.2).

Disk type	Limit P_S/P_E
Conventional flat	0.5
Conventional curved	0.7
Thick with inverse curve	0.9

Table 8.2. *Operating ratio*

8.6.3. *Relationship between pressure and flow*

We have seen that, upon rupture, the open cross-section immediately reaches its maximum value. The area of this cross-section is provided by the manufacturer, accounting for the presence of accessories and elements that could reduce the path open to fluid.

The relationship linking rupture pressure and flowrate for these disks is analogous to that established for safety valves.

1) Liquids:

The formula is the same as that for safety valves:

$$W = 0.62A\sqrt{2\rho P_E}$$

2) Gases:

We also use the formula for valves but introduce, as a precaution, as with liquids, restriction coefficient 0.62 of the fluid path:

$$W = 0.62\ A\ C\ P_1\sqrt{\frac{M}{Z\ R\ T_1}}$$

P_1 is the upstream pressure taken between the maximum operation pressure P_{MS} and the calculated pressure P_C of the protected device. If P_{DS} is the downstream pressure prior to rupture, we have:

$$P_1 = P_E + P_{DS}$$

Coefficient C is given by:

– non-critical flow:

$$C = r^{1/\gamma}\sqrt{\frac{2}{\gamma-1}\left[1-r^{\frac{\gamma-1}{\gamma}}\right]}$$

– critical flow:

$$C = \sqrt{\gamma\left(\frac{2}{\gamma+1}\right)^{\frac{\gamma+1}{\gamma-1}}}$$

EXAMPLE 8.7.–

The product is helium.

$\gamma = 1.66 \qquad P_1 = 1.24.10^6 \text{ Pa} \quad A = 0.00696 \text{ m}^2$

$M = 4 \qquad T_1 = 316.5 \text{ K}$

$$C = \sqrt{1.66\left(\frac{2}{2.66}\right)^{\frac{2.66}{0.66}}} = 0.725$$

$$W = 0.62\times 0.00696\times 0.725\times 1.24.10^6\sqrt{\frac{4}{8314\times 316.5}}$$

$$W = 4.78 \text{ kg.s}^{-1}$$

8.6.4. *Conventional disks*

The most straightforward design consists of a flat disk with low thickness. At DN 15, P_E can reach 40 bars and, at DN 500, be as low as 50 mbar. The operating ratio does not exceed 50%. The flat disk can be reinforced by a grill to withstand possible negative pressure.

If a value of P_E is required with several hundred bars, disk wall can be thin, but its surface curved with a concave turned toward the higher pressure value. Such disks can withstand an operating ratio of 70%. The vacuum resistance of curved disks is improved by the addition of a spherical cap on the central part of the disk. Thus, its curve is not in danger of being inversed (since it cannot be pressed down), if negative pressure governs in the device. In addition, disks reinforced in this way are less sensitive to pressure vibrations as, by retaining their shape, the metal wears less.

Conventional disks are made of Al, Ni, Monel, Inconel and stainless steel. They can be protected on one side by a sprayed-on coat of PTFE, CPV or epoxy resin.

8.6.5. *Composite disks*

These are made up of two overlapping thin sheets stuck against each other. They are vacuum resistant without the need for reinforcement:

– one sheet of a noble alloy ensures the greater part of mechanical resistance;

– the other sheet ensures corrosion protection and improves the seal while participating in mechanical resistance if it is made of a metal alloy, but can simply be plastic (essentially PTFE).

The possible range of P_E is wide (from 0.1 bar at DN 600 to 400 bar at DN 25) with improved protection against possible increases in temperature.

8.6.6. *Thick disks*

These disks are used in high-pressure polymerization installations in which the pressure can be in the order of 2,000 bar.

The disk surface is curved and, on its convex surface, has grooves that precut the pieces resulting from rupture. This helps avoid small fragmentation, which is a hazard for personnel working on it.

The concave side is polished to avoid polymer deposits and can also be coated with a protective plastic material like PTFE.

The thickness of these disks is at least three times greater than that would be required for a conventional disk of the same rupture pressure. The corrosion resistance is better as its lifespan is longer, so that the operating ratio reaches 0.9.

Nonetheless, if a thick disk must tolerate a counter-pressure greater than P_E without buckling, the surface curve must be inverse to the conventional shape. However, an inversely curved disk, grooved on the convex side, would bulge for a differential pressure in the order of 0.5 P_E. Thus, on both sides, inverting the curve can occur without disk rupture and to the detriment of rupture pressure reliability due to the prematurely fatigued metal. On inversing its curve, this disk hits one or several cutting blades, provoking rupture. However, the rupture pressure of such disks typically does not exceed a hundred bars.

While the conventional curve disks can be used with liquids without any particular precautions, the thick inverse curved disks require the presence of a small volume of gas in the liquid space.

8.6.7. *Graphite disks*

These are made of graphite impregnated with resin. This material has high resistance to chlorine products. The range available for P_E is approximately:

$$1\text{ bar} < P_E < 20\text{ bar} \qquad \text{at } \pm 5\%$$

Inversely, the minimum P_E can be as low as 30 mbar with DN higher by several hundred millimeters.

8.6.8. *Rupture indicators*

Three physical phenomena may be used:

– the movement of a metallic tape touching the center of the disk;

– to break with an electrical wire;

– photoelectric cell associated with the optical fiber.

In addition, the rupture of a disk must, in certain cases, initiate a series of safety features.

8.6.9. *Association of a disk with a safety valve*

Valves are fragile devices that are always sensitive to possible corrosion form the fluid of the protected device. By inserting a rupture disk between the valve and the device, we can maintain a fluid inoffensive to the plug and its seat. In addition, the presence of the disk prevents leaks that could occur through the valve. With this system, only the disk and not the valve would need to be made of high-quality materials, which help minimize costs.

A vent or pressure switch should be fitted between the disk and the valve to avoid or signal the accumulation of any pressure due to deterioration or cracking of the disk. Indeed, a cracked disk does not break completely and can significantly reduce the flow escaping from the valve.

Concerning this disk and valve association, the stacking of pressure is as follows:

$$P_S \leq P_{MS} \leq P_E - \Delta P_{EI}$$

And:

$$P_E + \Delta P_{ES} \leq P_T \leq P_C$$

P_S: operating pressure of the device

P_{MS}: maximum operating pressure of the device

P_T: valve set pressure

P_C: calculated pressure of the device.

If, as we described above, a disk is inserted upstream of a safety valve, we must accept *a priori* that the capacity of this valve is reduced by 20% since, on rupture, fragments of the disk can cause a partial obstruction of the cross-section open to the fluid path.

Of course, it is possible nonetheless to protect a valve from a corrosive atmosphere (exhaust side) by means of a flat disk with low P_E.

9

Breathing, Inerting, Gas Losses and Circulation between Reservoirs, Tanks and Vats

9.1. Breather valve specifications: reservoir filling ratio limitations

9.1.1. *Types of temperature fluctuations*

Fluctuations of the ambient temperature come mostly from two causes:

1) Daily variations between night and day:

In arid regions, the amplitude of daily variations can reach 40°C, while in Europe, it seldom exceeds 10°C, for instance, 10°C at night and 20°C in day.

However, it is well advised to move to the higher range of variation due to it corresponding to higher pressure vapor and the maximum values of losses. Thus, in bulletin API 2516, the extreme values retained are 30 and 60°C.

The liquid contained in reservoirs is typically at a temperature close to the lower limit, which is that of the cooling water used in installations.

In daytime, the free surface of the liquid is heated 5–6 times more slowly than the surrounding atmosphere, so if the atmosphere heats by 30°C, the liquid surface only heats by 5–6°C. For example, in bulletin API 2516, it is accepted that the superficial temperature of stored liquids varies between 32 and 38°C.

2) Succession of rainy and clear weather:

When rain follows a clear weather, cooling seldom exceeds 10°C.

The heat flow density transmitted to the lining of a non-insulated reservoir is maximum when rain follows a clear weather. With a convection coefficient in the order of 8 $W.m^{-2}.°C^{-1}$ and a difference of 10°C, the flow density is 80 $W.m^{-2}$.

On the other hand, heating by solar radiation is less intense and only in the order of 40 $W.m^{-2}$.

3) The temperature variation between night and day determines the set pressure of the breather safety valves. It is the succession of rainy and clear weather that is used to determine the gas flow through the valve.

9.1.2. *Set pressure of exhalation valve*

The set pressure P_M must be sufficiently high so that there is no exhaust when the reservoir contents are heated.

The mass of trapped air remains constant along with the volume of the mixed air and vapor.

According to the law of perfect gases applied to air:

$$\frac{nR}{V} = \text{cste} = \frac{P_{air}}{T} = \frac{P_{Tot} - \pi}{T}$$

That is:

$$\frac{P_M - \pi_M}{t_M + 273} = \frac{P_m - \pi_m}{t_m + 273}$$

t_M and t_m are the maximum and minimum ambient temperatures.

π_M and π_m are the pressure values of the saturation vapor on the free surface of the liquid at the maximum and minimum temperatures, respectively.

Typically, P_m is established by the mechanical resistance of the reservoir to negative pressure (P_m = 1.008 bar) and the previous relationship provides P_M. P_m is the set pressure of the intake valve.

EXAMPLE 9.1.–

$$P_m = 1.008 \text{ bar}$$

$$\pi_m = \pi\,(32°) = 0.84 \text{ bar} \quad \pi_M = \pi\,(38°) = 1 \text{ bar}$$

$$t_m = 30°C \qquad t_M = 60°C$$

$$P_M = 1 + (1.008 - 0.84)\left[\frac{60+273}{30+273}\right]$$

$$P_M = 1.287 \text{ bar abs.}$$

NOTE.–

If the reservoir is located *inside* rather than *outside*, the amplitude of temperature variations can be divided by approximately 3 or 5 depending on whether it is a cladded structure or a structure closed by walls.

NOTE (Extraction possible without breathing).–

Let V be the initial gas volume in the reservoir at P_m.

We withdraw ΔV of liquid, which provokes the entry of ΔV of air at P_m.

Volume ΔV of air will saturate, provoking the compression of the whole gas from pressure P_m to pressure P_M.

At P_M, the volume occupied by the vapor released is:

$$\Delta V \frac{\pi}{P_M}$$

This volume occupies the space freed by the compression of V + ΔV from P_m to P_M.

$$(V+\Delta V)\left(1-\frac{P_m}{P_M}\right)=\Delta V\frac{\pi}{P_M}$$

Hence:

$$\frac{\Delta V}{V}=\frac{P_M-P_m}{\pi+P_m-P_M}$$

EXAMPLE 9.2.–

P_M-P_m = 0.03 bar

$\pi = 0.2$ bar

$$\frac{\Delta V}{V}=\frac{0.03}{0.2-0.03}=0.176=17.6\%$$

9.1.3. *Range of filling ratio for operations*

While, as we have seen above, losses due to fluctuations in the exterior temperature can be avoided, operating losses (due to replacing the reservoir contents) cannot be entirely avoided.

However, as we will see, the greater the $P_M - P_m$ difference of the inlet and outlet valves, the larger the possible amplitude of level variations without breathing.

Moreover, we will show that the possible variation in liquid level without breathing is more significant with higher volume blanketing. This will show the interest to be found in, for example, not filling a buffer reservoir (within a process) to more than 60% of its height.

If V_T is the total volume of the reservoir and R the mean filling ratio, volume V_G of the blanketing is:

$$V_G=(1-R)V_T$$

The pressure governing in the reservoir is P_0, so:

$$P_m < P_0 < P_M$$

Production limitations can provoke filling variations with the maximum amplitude equal to $\pm\ \Delta V$ and, during an increase in the level, the pressure should remain below P_M and, on a decrease in the level, it should remain above P_m. In other words:

On filling, we must have: $V_0 - \Delta V > P_0 V_0 / P_M$

On emptying, we must have: $V_0 + \Delta V < P_0 V_0 / P_m$

That is:

– for partial filling: $$R < 1\ \Delta \frac{\Delta V/V_T}{1 - \dfrac{P_0}{P_M}}$$

– for partial filling: $$R < 1\ \Delta \frac{\Delta V/V_T}{\dfrac{P_0}{P_M} - 1}$$

EXAMPLE 9.3.–

Let us suppose that the inevitable flow fluctuations are such that (where V_T is the total reservoir volume):

$$\Delta V/V_T = 0.1$$

Also supposing that:

$$\frac{P_0}{P_m} = 1.15 \quad \text{and} \quad \frac{P_0}{P_M} = 0.85$$

Incidentally, we should note that:

$$\frac{P_M}{P_m} = \frac{1.15}{0.85} = 1.353$$

Hence, in this case:

$$1-\frac{P_0}{P_M}=0.15=\frac{P_0}{P_m}-1$$

The filling rate R must then be limited as:

$$R<1-\frac{0.1}{0.15}=0.33$$

The reservoir should only be filled to 1/3 of its total volume.

NOTE.–

We have seen that in order to compensate temperature fluctuations, we should set the outlet valve accordingly.

9.1.4. *Outlet valve flow*

The following four criteria need to be followed:

– 1: The gas volume flowrate crossing the valve in both directions must be at least equal to the liquid flow volume for filling and emptying. These operations often occur in 10 min for tanks that are not too large in volume.

– 2: The gas flowrate from the valve must at least correspond to a temperature variation of 55°C per hour of an ideal gas mixture which completely fills the tank. This speed of temperature variation is compliant to norm API 2000:

$$\frac{1}{V}\frac{dV}{d\tau}=\alpha\frac{dt}{d\tau}=\frac{55}{273}=0.2$$

The real flow in $m^3.h^{-1}$ is consequently:

$$Q=0.2V_T \quad (V_T\text{: total volume of the reservoir})$$

– 3: During the rapid variation of the ambient temperature, the thermal power transmitted to the tank is (in Watts) equal to 80 S, where S is the exposed surface. Accordingly, we can estimate the exposed surface by:

$$S = 4.7V^{2/3} \qquad (m^2)$$

V: Useful volume of the reservoir (m^3).

Where L ($J.kg^{-1}$) is the latent heat of the product, the evaporated mass is:

$$W_E = \frac{80 \times 4.7 \times V^{2/3} \times 3600}{L} \qquad (kg.h^{-1})$$

And the total flow in volume is:

$$Q_N = \frac{W_E}{M} \times \frac{P_M}{\pi} \times 22.42 \qquad (Nm^3.h^{t1})$$

M: molar mass of product ($kg.kmol^{-1}$).

– 4: The product is stored in a place where the inflammable products are stored, the thermal power received by the tank from the neighboring flames during a fire is:

$$43.17\ A^{0.82} \qquad (kW)$$

A is the surface of the reservoir exposed to the flames and expressed in m^2.

By dividing the latent heat L (e, $kJ.kg^{-1}$) of the contained liquid, we obtain the evaporated flow in mass:

$$W_E = \frac{43.17}{L} A^{0.82} \times 3600 \qquad \left(kg.h^{-1}\right)$$

The flow that subsequently has to pass through the valve is:

$$Q_N = \frac{W_E}{M} \times 22.42 \qquad (Nm^3.h^{t1})$$

If t_E is the ebullition temperature at P_M, the real flow is:

$$Q=\frac{W_E}{M}\times 22.4\times\frac{t_E+273}{273}\times\frac{1.013}{P_M} \qquad (m^3.h^{t1})$$

NOTE.–

The calculation of surface A exposed to the fire is described in the applicable rules that must be consulted:

– for vertical reservoirs: if the height is less than or equal to 10 m, surface A comprises the lateral surface and top cover. If height is more than 10 m, we only take lateral height into account for a height of 10 m;

– for horizontal reservoirs, we take 75% of the total surface;

– for spheres and ellipsoids or spheroids, we take 55% of the total surface or the exterior surface up to a height of 10 m (we take the greater of the two values).

NOTE.–

If the product is inflammable, a flame-arrester must be installed on the valve exhaust (and on the inlet) if the flash point of the product is below 38°C (consult rules) in order to prevent the propagation of outside fire to the inside of the reservoir and vice versa.

EXAMPLE 9.4.–

Reservoir volumes: 632 m^3 and 15,800 m^3

$t_E = 50°C$ $M = 61$ $\pi = 0.7$ bar $P_M = 1.1$ bar abs. L = 334 kJ:kg

No. 1:

$632 \times 4 = 2.528\ m^3/h$ and $15{,}800/2 = 7{,}900\ m^3.h^{-1}$

No. 2:

$0.2 \times 632 = 126\ m^3/h$ and $15{,}800 \times 0.2 = 3{,}160\ m^3.h^{-1}$

No. 3:

V	632	15,800	
S	345	2,940	
W_E	297	2,535	
Q_N	171	1,464	(at 0°C and 1 atm.)
Q	169	1,446	(at 20°C and 1.1 bar)

No. 4:

Let us equate A with S for the small reservoir and A with S/2 for the larger reservoir.

V	632	15,800	
S	345	1,470	
W_E	56,000	184,000	
Q_N	20,582	67,627	(at 0°C and 1 atm.)
Q	20,335	66,815	(at 20°C and 1.1 bar)

So, according to Table 1 of norm API 2000:

V	632	15,800
S	113	1,700

And finally the Q values (real $m^3.h^{-1}$).

V	632	15,800
1	2,528	7,900
2	126	3,160
3	169	1,446
API 2000	113	1,700
4	20,335	66,815

In this example, exploitation and fire are determinant.

9.1.5. *Pressure reducers and overflow valves*

Traditional breather valves are only intended for over-pressures below 0.1 or 0.2 bar and negative pressure below 5–10 mbar. These ranges are suitable if the tank's operating pressure is close to the atmospheric pressure and if atmospheric air can be accepted into the reservoir.

On the other hand, if the gas atmosphere must be made up of an inert gas (nitrogen for instance), these gases are available at high pressure and must consequently be made to cross one or two reducers in series, with the consecutive exhaust of the level increase occurring by means of an overpressure valve. The operating principle of these devices is represented in Figure 9.1.

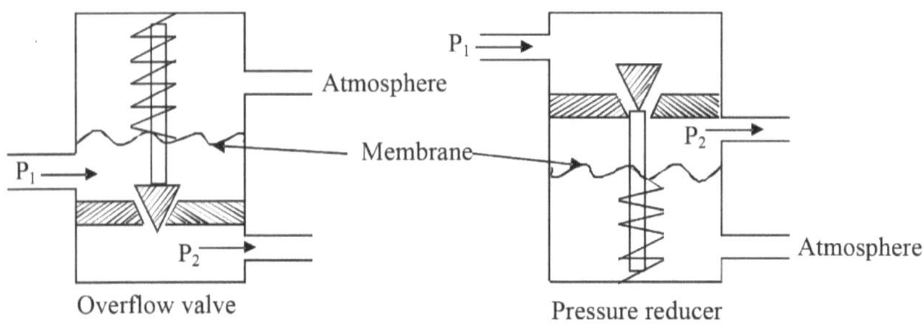

Figure 9.1. *Difference between overflow valve and pressure reducer*

For example, if pressure P_2 upstream of the reducer is too high, it acts on the membrane which presses on the needle, reducing the arriving gas. On the other hand, if P_2 is too low, the spring lifts the needle, which increases the gas flow.

The overpressure valve regulates upstream pressure rather than downstream pressure.

Thus, we can keep a blanket of inert gas up to a pressure in the order of 10 bar.

9.1.6. *Possible mounting for use of inert gas under pressure*

An improved version would consist of allowing not only two pressure reducers in series, but also:

– an upstream reducer on the high pressure;

– two parallel reducers downstream of the upstream reducer, of which one, with low capacity, performs fine-tuning, while the other, with high capacity, allows for the fast arrival of inert gas at the beginning of an operation. The high-capacity reducer regulates the existing pressure upstream of the reducer used for fine-tuning.

We refer to Figure 9.2.

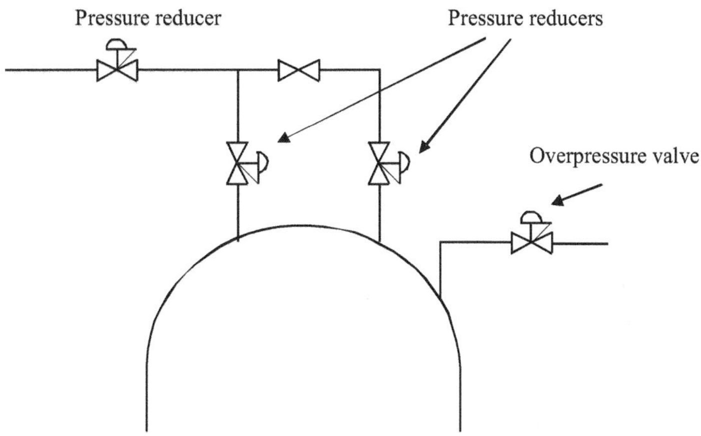

Figure 9.2. *Inert volume*

9.2. Assessment of losses to the atmosphere

9.2.1. *Preliminaries*

Apart from the study we will perform on losses due to overheating, we assume here that the reservoirs are used at a constant temperature.

During the course of operations, the renewal of reservoir contents occurs periodically. In other words, each time the level goes down (withdrawal), it is followed by an equivalent increase in the level (filling). This cycle corresponds to a renewal, that is:

– partial if the ΔV volume concerned is less than the useful volume RV_T of the reservoir, where R is the filling rate;

– total if $\Delta V = RV_T$ where V_T is the total volume.

The tanks, and vats that we consider here are typically equipped with a breather valve comprising two plugs:

– an outlet plug opening for internal over-pressurization (relative to the external pressure P_a) equal to $P_M - P_a$. In general, the values of this over-pressurization are between 25 bar and 200–300 mbar;

– an inlet plug opening for internal negative pressure (relative to the outside pressure P_a) equal to $P_a - P_m$. In general, the values of this negative pressure are between 2 and 5 mbar, only if we need to avoid an expensive tank design. Indeed, resistance to negative pressure requires significant reinforcements.

9.2.2. *Possible scenarios*

1) On storing a raw material, filling is fast and withdrawal is slow (at the rhythm of manufacturing requirements).

2) On storing an end product, filling is slow (manufacturing rhythm), and withdrawal is fast.

3) On use of a buffer tank within a process, filling variations are limited.

9.2.3. *Raw material storage*

The tank, initially empty of liquid but containing a mixture of air and vapor, is filled quickly. The mixture is at total pressure P_m and is saturated with the product, that is, the partial pressure of the product is equal to its vapor pressure $\pi(t)$ at set temperature t.

On filling, a fraction τ of the volume RV_T of the liquid is used to make the gas phase go from pressure P_m to pressure P_M.

Initially, we can write:

$$V_T P_m = n_o rT$$

r: perfect gas constant (8,314 $J.kmol.^{-1}K^{-1}$)

T: absolute temperature (K).

The conservation of the number n_o of kilomoles of gas is written as:

$$V_T P_m = (V_T - \tau R V_T) P_M$$

Hence:

$$\tau = \frac{1}{R}\left(1 - \frac{P_m}{P_M}\right)$$

The liquid volume $(1 - \tau)\ RV_T$ takes the place of an equivalent volume of saturated gas at the total pressure P_M. Where n_e is the total number of kilomoles evacuated to the atmosphere:

$$n_e rT = \pi(1-\tau)RV_T$$

The mass of product lost is:

$$Mn_e = \frac{\pi(1-\tau)RV_T}{rT}$$

M: Molar mass of product ($kg.kmol^{-1}$)

The mass lost, applied to the useful volume RV_T, is:

$$\frac{\Delta m}{RV_T} = \frac{Mn_e}{RV_T} = \frac{M\pi(1-\tau)}{rT}$$

9.2.4. *Product storage*

The reservoir is filled slowly at the manufacturing rhythm but emptied quickly and the air replaces the liquid volume RV_T.

The partial pressure of the vapor in the empty reservoir is:

$$P_{V0} = \pi(1-R)$$

Saturation is reached when:

$$P_V = \pi$$

When vaporization occurs, the number of kilomoles dn_V vaporized corresponds to an increased dP_V:

$$dn_V = \frac{V_T dP_V}{rT}$$

In order for the total pressure to go from P_m to P_M, we must vaporize

$$dn_V = \frac{V_T \left(P_M - P_m\right)}{rT}$$

Which is simply:

$$\Delta P_V = P_M - P_m \quad \text{and} \quad P_{V1} = P_{V0} + \Delta P_V = \pi\left(1-R\right) + P_M - P_m$$

Vaporization then continues with the valve open, that is at constant pressure P_M.

$$n_V = \frac{P_M dV}{rT}$$

The vaporized product has two effects:

– increasing the pressure of the product's partial vapor:

$$\frac{V_T dV_V}{rT}$$

– compensating vapor expulsion:

$$\frac{P_V dV}{rT}$$

Hence:

$$P_M dV = P_V dV + V_T dP_V$$

$$\frac{dV}{V_T} = \frac{dP_V}{P_M - P_V}$$

The loss of vapor to the atmosphere is:

$$n_e = \frac{1}{rT}\int P_V dV = \frac{V_T}{RT}\int_{P_{V1}}^{\pi} \frac{P_V dP_V}{P_M - P_V}$$

Therefore, in kg.m^{-3} of liquid:

$$\frac{\Delta m}{RV_T} = \frac{M}{rTR}\left[P_M \mathrm{Ln}\left[\frac{P_M - P_{V1}}{P_M - \pi}\right] - (\pi - P_{V1})\right]$$

EXAMPLES 9.5.–

$R = 0.6$	$\pi = 0.7$ bar
$t = 20°C = 293$ K	$P_M = 1.2$ bar
$M = 61$ kg.kmol^{-1}	$P_m = 1$ bar

Raw material storage:

$$\tau = \frac{1}{0.6}\left(1 - \frac{1}{1.2}\right) = 0.28$$

$$\frac{\Delta m}{RV_T} = \frac{61 \times 0.7.10^5 \times (1 - 0.28)}{8314 \times 293} = 1.26 \text{ kg.m}^{-3}$$

End product storage:

$$P_{V0} = 0.7.10^5(1 - 0.6) = 0.28.10^5 \text{ Pa}$$

$$P_{V1} = 0.28 + 0.2 = 0.48 \text{ bar} = 0.48.10^5 \text{ Pa}$$

$$\frac{\Delta m}{RV_T} = \frac{61}{8314 \times 293 \times 0.6}\left[1.2.10^5 \mathrm{Ln}\left[\frac{1.2 - 0.48}{1.2 - 0.7}\right] - (0.7 - 0.48).10^5\right]$$

$$\frac{\Delta m}{RV_T} = 0.90 \text{ kg.m}^{-3}$$

These losses are low, but it can be unacceptable if the product or raw material is toxic.

If the product or raw material is inflammable, we can provide an option for nitrogen entry to replace the air entry. A gas blanket of a different gas to air can also be provided if the liquid is toxic.

NOTE.–

In cases where a buffer tank is used during a process, the predictions are more complex, and depend on both the amplitude and the speed and frequency of changes in level and pressure inside the reservoir.

Cases need to be studied on an individual basis. The ideal situation is that in which the internal pressure remains in the range of $P_M - P_m$. Where this is the case, there are no exchanges with the ambient atmosphere.

9.2.5. *Losses on overheating*

At set pressure P_M of the outlet valve and with conditions t_M and π_M, the volume of the gas blanket is V_0, and the law of perfect gases applied to the air present is written as:

$$V_0 (P_M - \pi_M) = n_{air} r (t_M + 273)$$

On heating:

$$t > t_M \quad \text{and} \quad \pi > \pi_M$$

Hence:

$$V_0 + \Delta V = \frac{V_0 (P_M - \pi_M)}{(t_M + 273)} \times \frac{(t + 273)}{(P_M - \pi)}$$

Thus, the gas blanket is dilated and the loss escaping corresponds to volume ΔV at t, P_M and π.

The concentration c of the product of the gas mixture is (in kg per real m^3):

$$c = \frac{M\pi.10^5}{8314\times(t+273)}$$

M: molar mass of product ($kg.kmol^{-1}$).

Finally:

$$\frac{\Delta V}{V_0} = \left[\left[\frac{t+273}{t_M+273}\right]\times\left[\frac{P_M-\pi_M}{P_M-\pi}\right]-1\right]$$

Hence:

$$\frac{\Delta M}{V_0} = \frac{M\pi.10^5}{8314(t+273)}\left[\left[\frac{t+273}{t_M+273}\right]\times\left[\frac{P_M-\pi_M}{P_M-\pi}\right]-1\right]$$

EXAMPLE 9.6.–

$P_M = 1.1$ bar	$M = 63\ kg.kmol^{-1}$
$\pi_M = 0.7$ bar	$\pi = 0.8$ bar
$t_M = 40°C$	$t = 50°C$

$$\frac{\Delta M}{V_0} = \frac{63\times0.8.10^5}{8314.3\times323}\left[\frac{323}{313}\times\frac{0.4}{0.3}-1\right] = 0.7\ kg/m^{-3}$$

If the gas blanket occupies 25% of the reservoir volume, the loss applied to the liquid present is:

$$\frac{0.7\times0.25}{(1-0.25)} = 0.233\ kg.m^{-3}$$

And if the density of the liquid is 600 $kg.m^{-3}$, the weight loss during this heating is:

$$0.233/600 = 0.00039 = 0.04\%$$

9.3. Circulation of liquid between reservoirs

9.3.1. *Transfer by gravity from one reservoir to another*

In the source tank, the liquid level goes from value H_{s0} to value H_s and, in the use tank, the liquid goes from value H_{u0} to H_u. The conservation of volume Ω of transferred liquid is written as:

$$\Omega = A_s\left(H_{s0} - H_s\right) = A_u\left(H_u - H_{u0}\right)$$

Or rather:

$$\Omega = \frac{H_{s0} - H_s}{\frac{1}{A_s}} = \frac{H_u - H_{u0}}{\frac{1}{A_u}} = \frac{\left(H_{s0} - H_{u0}\right) - \left(H_s - H_u\right)}{\frac{1}{A_s} + \frac{1}{A_u}}$$

Write out:

$$\frac{1}{\bar{A}} = \frac{1}{A_s} + \frac{1}{A_u}$$

Furthermore, the reduction of the hydrostatic charge is:

$$\Delta H = \left(H_{s0} - H_{u0}\right) - \left(H_s - H_u\right) \qquad \Delta H > 0$$

We observe that the volume transferred is simply expressed by:

$$\Omega = \bar{A}\Delta H$$

And again:

$$Q = \frac{d\Omega}{d\tau} = -\bar{A}\frac{dH}{d\tau} \qquad \frac{dH}{d\tau} < 0$$

We may well draw a distinction according to whether flow occurs in laminar or turbulent regime.

1) Laminar regime:

The pressure drop in a pipe is:

$$\Delta P = \frac{32\mu L V}{D^2} = \frac{128\mu L Q}{\pi D^4}$$

ΔP: pressure drop (Pa)

μ: viscosity of liquid (Pa.s)

L: pipe length (m)

D: interior diameter of the pipe (m)

V: velocity of the liquid ($m.s^{-1}$)

Q: flowrate in volume ($m^3.s^{-1}$)

With the action of the hydrostatic height H:

$$H = \frac{128\mu L\bar{A}}{\rho g \pi D^4}\frac{dH}{d\tau}$$

Integrating:

$$\tau = K\,\mathrm{Ln}\frac{H_0}{H} \qquad \text{with}: K = \frac{128\mu L\bar{A}}{\rho g \pi D^4}$$

We should recall that the flow regime is laminar if the Reynolds number is less than 2,100.

2) Turbulent regime:

According to the notations above, the pressure drop in a pipe is:

$$\Delta P = f\frac{L}{D}\frac{\rho V^2}{2} = \frac{8fL}{\pi^2 D^5}\rho Q^2$$

f: friction coefficient (in the order of 0.03)

ρ: density of the liquid ($kg.m^{-3}$).

With the hydrostatic pressure ρgH:

$$H = \frac{8fL}{g\pi^2 D^5} Q^2 = K^2 \left(\bar{A} \frac{dH}{d\tau} \right)^2$$

Or rather:

$$d\tau = -K\,\bar{A} \frac{dH}{\sqrt{H}}$$

That is, integrating:

$$\tau = 2K\,\bar{A} \left(\sqrt{H_0} - \sqrt{H} \right)$$

With:

$$K = \sqrt{\frac{8fL}{g\pi^2 D^5}} \qquad \left(s.m^{-2.5} \right)$$

EXAMPLES 9.7.–

$A_s = 78.5\ m^2$ $H_0 = 25\ m$ $\rho = 900\ kg.m^{-3}$

$A_u = 19.6\ m^2$ $H = 4\ m$

$D = 0.15\ m$ $L = 100\ m$

$\bar{A} = 15.6\ m^2$

1) Viscous product. Laminar regime:

$\mu = 0.5\ Pa.s$

$$K = \left[\frac{128 \times 0.5 \times 100 \times 15.6}{900 \times 9.81 \times \pi \times 0.15^4} \right]^{0.5} = 7110$$

The duration of the operation is:

$$\tau = 7110 \text{ Ln}\frac{25}{4} = 13030 \text{ s}$$

$$\tau = 3 \text{ h}37 \text{ mn}10 \text{ s}$$

2) Very fluid product. Turbulent regime:

$$\mu = 0.001 \text{ Pa.s}$$

$$K = \sqrt{\frac{8 \times 0.03 \times 100}{9.81 \times \pi \times 0.15^5}} = 57.13$$

$$\tau = 2 \times 57.13 \times 15.6\left(\sqrt{25} - \sqrt{4}\right) = 5347$$

$$\tau = 1 \text{ h}29 \text{ mn}70 \text{ s}$$

NOTE (Verifications).–

At the start of the laminar regime, the velocity is:

$$V = \frac{900 \times 9.81 \times 25 \times 0.15^2}{32 \times 0.5 \times 100} = 3.1 \text{ m.s}^{-1}$$

$$Re = \frac{3.1 \times 0.15 \times 900}{0.5} = 837 < 2100$$

At the end of the turbulent regime, the velocity is:

$$V = \sqrt{\frac{2 \times 9.81 \times 4 \times 0.15}{0.03 \times 100}} = 1.98 \text{ m.s}^{-1}$$

$$Re = \frac{1.98 \times 0.15 \times 900}{0.001} = 267.300 >> 2100$$

10

Flow in Pipes: Rarified Gas, Non-Newtonian Liquids, Events, Gas–Liquid Flow

10.1. Rarified gas

10.1.1. *Viscosity of gas under low pressure*

In order to calculate this viscosity, Bromley and Wilke [BRO 51] propose the following expression (their equation 19):

$$\eta = \frac{4,23.10^{-4} M^{1/2} P_c^{2/3}}{T_c^{1/6}} f\left(\frac{kT}{\varepsilon}\right)$$

Table I of their work gives the value of the ε/k ratio in Kelvin for 45 gases that are used widely in the industry.

Table II gives the values of function $f\left(\frac{kT}{\varepsilon}\right)$. However, Bromsley and Wilke simply propose:

$$f\left(\frac{kT}{\varepsilon}\right) = 08788\left(\frac{kT}{\varepsilon}\right) \quad \text{when} \quad 10 < \frac{kT}{\varepsilon} < 400$$

M: molar mass (g.(gmol)$^{-1}$ or rather kg.kmol^{-1})

P_c: critical pressure (atm.)

T_c: critical temperature (K)

η: viscosity of gas (cP (1 centipoise = 10^{-3} Pa.s))

In the work of Wilke [WIL 50], we find an expression for the viscosity of gas mixes at low pressure.

EXAMPLE 10.1.–

Viscosity of nitrogen at 20°C = 293 K

According to Table I: $\varepsilon/k = 91.46$

$$\frac{kT}{\varepsilon} = \frac{293}{91,46} = 3,20$$

In Table II, we find:

$$f\left(\frac{kT}{\varepsilon}\right) = 1,7573$$

According to the Appendix:

$M = 28$ kg.kmol^{-1} $\quad$ $P_c = 33,9$ atm. $\quad$ $T_c = 127$ K

$$\eta = \frac{4.23.10^{-4} \times 28^{1/2} \times 33.9^{0.666} \times 1.7573}{127^{1/6}}$$

$$\eta = 0.0183 \text{ cp} = 0.183.10^{-4} \text{ Pa.s}$$

10.1.2. *Quadratic mean velocity*

$$\overline{v} = \left(\frac{8RT}{\pi M}\right)^{1/2}$$

R: perfect gas constant (8.314 J.kmol^{-1}.K^{-1})

T: absolute temperature (K)

M: molar mass (kg.kmol^{-1})

EXAMPLE 10.2.–

$$T = 293 \text{ K} \qquad M = 29 \text{ kg.kmol}^{-1}$$

$$\overline{v} = \left(\frac{8 \times 8314 \times 300}{\pi \times 29}\right)^{1/2} = 467 \text{ m.s}^{-1}$$

10.1.3. *Mean free path*

$$\lambda = \frac{\alpha\eta}{cM\overline{v}}$$

α: adimensional coefficient

c: gas concentration (kmol.m^{-3})

M: molar mass (kg.kmol^{-1})

η: gas viscosity (Pa.s = kg.s^{-1}.m^{-1}).

According to Scott and Dullien [SCO 62], coefficient α is equal to 2.

10.1.4. *Gas flow (low-pressure pipe)*

Under low pressure, all gases are eminently compressible. In this situation, Brun *et al.* [BRU 68] give the mass flow expression for a gas going from pressure P_1 to the lower pressure P_2 in pipe length L.

$$W = \frac{\pi}{16\eta} \cdot \frac{M}{RT} \cdot \frac{r^4}{L}\left(P_1^2 - P_2^2\right)$$

η: gas viscosity (Pa.s)

W: mass flowrate (kg.s^{-1})

R: perfect gas constant (8.314 J.kmol^{-1}.K^{-1})

T: absolute temperature (K)

M: molar mass of gas ($kg.kmol^{-1}$)

r: radius of the pipe with circular cross-section (m)

P_1 and P_2: upstream and downstream pressures of pipe (Pa).

Let us introduce the mean pressure P_m so that:

$$P_m = \frac{1}{2}(P_1 + P_2)$$

Let us define parameter S_o, so that:

$$S_o = \frac{\pi r^4}{8\eta L} P_m$$

Building on the work of Knudsen [KNU 09], Brown *et al.* [BRO 46] correct the expression of Brun *et al.* [BRU 68] (that is also known as the Poiseuille equation). To this end, for pressures less than 1 Torr, they write:

$$S = FS_o$$

That is:

$$\frac{RTW}{M(P_1 - P_2)} = S = \frac{F\pi r^4 P_m}{8\eta L}$$

Recall that 1 Torr = 1 mm column of mercury.

Brown *et al.* [BRO 46] give four curves to calculate F:

– their Figures 1 and 2 for glass capillaries

– their Figure 3 for copper tubes

– their Figure 4 for steel tubes.

In these figures, the horizontal value is parameter:

$$X = \frac{\eta}{P_m r \sqrt{\frac{M}{RT}}}$$

Parameter X is adimensional.

NOTE.–

Brown *et al.* [BRO 46] complicate matters by introducing the micrometer of mercury as a unit of pressure, which leads them to introduce the coefficient $g_c = 1.33$. Effectively:

$$1\mu\text{m Hg} = \frac{981 \times 13.6 \times 1}{10000} = \frac{13341.6}{10000} = 1.33 \text{ barye}$$

The barye ($g.cm^{-1}.s^{-2}$) is the pressure unit of the C.G.S. system.

NOTE.–

According to Brown *et al.* [BRO 46] for a Reynolds number below 100 (which is the case of industrial pipes under vacuum), the resistance of 40 and 80 cm elbows can be neglected. A bellows valve is equivalent to a pipe whose length is five times its diameter.

NOTE.–

Scott and Dullien [SCO 62] propose an expression for gas flow in which all parameters can be obtained from the kinetic theory of gases. This is their equation 21. This method avoids the use of curves.

NOTE.–

A reader who wishes to learn more about the phenomenon of gas friction at very low pressure on a solid surface can refer to the work of Knudsen [KNU 09] and that of Millikan [MIL 23].

10.2. Consistent or plastic products

10.2.1. *Pipe flow laws*

Recalling the Poiseuille law for flow volume of a Newtonian liquid (see Brun *et al.* [BRU 68]):

$$Q = \frac{\pi(P_1 - P_2)r^4}{8\eta L}$$

Q: flow ($m^3.s^{-1}$)

P_1 and P_2: upstream and downstream pressure (Pa)

r: radius of the pipe with circular cross-section (m)

η: liquid viscosity (Pa.s)

L: pipe length (m).

If the product is non-Newtonian and non-plastic, the sheering force within the liquid is:

$$\tau = k\left(-\frac{du}{dy}\right)^n$$

u: local axial velocity at radial distance y

y: axis distance (from the pipe) from the point considered

τ: sheering force (Pa)

k: consistence coefficient (or, to be more concise: *consistence*)

n: rheotypic exponent: adimensional

Q: flowrate ($m^3.s^{-1}$)

Consistence is measured in $Pa.s^n$.

A calculation analogous to the one leading to the Poiseuille law gives the following result for the flowrate volume:

$$Q = \pi \left(\frac{P_1 - P_2}{2kL} \right)^{1/n} \left(\frac{n}{3n+1} \right) r^{\frac{3n+1}{n}}$$

Nature of the product	Temperature °C	Consistence k Pa.s	Rheotypical exponent
Pear compote (or apple)	25	0.05	0.38
Peach compote	25	0.05	0.34
Tomato paste (30% m.s.)	32	19	0.4
Crème fraiche (30% m.gr.)	3	1	1
Olive oil	20	0.09	1
Honey	25	6	1

Table 10.1. *Rheological properties of several products*

When kept in a fridge, honey develops crystals (glucose?) and is no longer Newtonian.

Therefore:

Pseudo-plastic where $n < 1$

Dilatant where $n > 1$

Newtonian where $n = 1$.

NOTE.–

Darby [DAR 81] addressed the problem of plastic fluid flow in the Bingham meaning. For such fluids, the flow is only possible if the shearing force does exceed a critical value.

10.3. Vents

10.3.1. *General*

There are two categories of vents:

– when a rupture disk breaks following an explosion inside a receptacle, the gases that cross the disk must be evacuated, not into the ambient workshop atmosphere but to the outside. Accordingly, we must provide a pipe connecting the rupture disk and a zone where the presence of the gases issued from the explosion is not problematic. This pipe is called the *explosion vent*;

– when receptacles have to be permanently kept at atmospheric pressure, but if the gas atmosphere of each receptacle above the liquid is a mixture of gas that is toxic for the environment, we must provide a pipe connecting the receptacle to a place where these gases have no toxic effects. We call such a pipe an *ambient vent*. This of course refers to the *ambience in the receptacle*.

10.3.2. *Explosion vents*

Singh [SIN 79] addressed the problem of these vents in detail. In particular, they require:

– an adimensional parameter characteristic of the recipient shape;

– the propagation speed of the fire that results from the explosion;

– the vent cross-section.

Depending on whether the over-pressurization resulting from the explosion is less than or more than 1 atm., the explosion is a deflagration or a detonation. Of course, the vent must be able to withstand the over-pressure.

NOTE.–

The pipe servicing the vent must:

– start from the rupture disk;

– end with a vertical segment fitted with a cover that opens when its weight is compensated by the gas pressure.

Singh [SIN 79] described the advantages and disadvantages of the weighted cover with regard to the rupture disk.

10.3.3. *Ambient vents*

We suppose that the material (solid or liquid) is located in a receptacle that is isolated from the atmosphere, but where pressure is close to atmospheric pressure. Dangerous gases (toxic, explosive) can be present in the receptacle. In such cases, we must evacuate these gases using a vent that leads them, where they are no longer dangerous.

Vents serving different devices all lead to a general collector at the end of which a fan takes the gases and sends them:

– to the atmosphere;

– to the place of treatment (a packed column or rather a pinching scrubber).

Typically, we cannot modulate the flow of dangerous output gases; however, ventilators have to operate with a flowrate fluctuating through precise limits. The solution consists of adding air to each vent.

To this end, each vent is fitted with an overturned funnel and the air enters as indicated in Figure 10.1.

Write out:

W_E: mass flowrate naturally emitted (kg.s^{-1})

W_A: mass flowrate of air entering the vent on output exit (kg.s^{-1})

W_R: mass flowrate in the vent (kg.s^{-1}).

We must have:

$$W_R = W_E + W_A$$

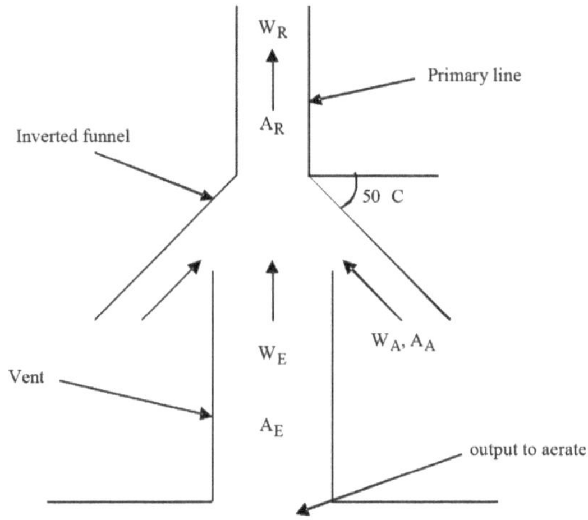

Figure 10.1. *Air entry on vent exit*

Also writing out:

A_E: cross-section of output exit tube (m^2)

A_A: cross-section of crown perpendicular to air flow W_A (m^2)

A_R: cross-section of primary line to collector (m^2)

$$x = \frac{W_A}{W_E} \qquad \tau = \frac{A_A}{A_E} \qquad r = \frac{A_R}{A_E} < 1 \tag{10.1}$$

$$\rho_R = \frac{W_A + W_E}{\dfrac{W_A}{\rho_A} + \dfrac{W_E}{\rho_E}} = \frac{1 + x}{\dfrac{1}{\rho_E} + \dfrac{x}{\rho_A}} \tag{10.2}$$

ρ_A and ρ_E: densities of ambient air and gas blanket ($kg.m^{-3}$)

The conservation of the momentum is written as:

$$W_R V_R = W_A V_A + W_E V_E \quad \text{avec} \quad W_R = W_A + W_E$$

The Vs are the velocities:

$$V = \frac{W}{\rho A}$$

Therefore:

$$\frac{(W_A + W_E)^2}{A_R \rho_R} = \frac{W_A^2}{\rho_A A_A} + \frac{W_E^2}{\rho_E A_E} \quad \text{or rather} \quad \frac{(1+x)^2}{r\,\rho_R} - \frac{1}{\rho_E} = \frac{x^2}{\tau\,\rho_A}$$

$$\tau = \frac{x^2 / \rho_A}{\dfrac{(1+x)^2}{r\,\rho_R} - \dfrac{1}{\rho_E}} \qquad [10.3]$$

where t_E is the temperature of the output emitted and $\pi(t_E)$ the saturating vapor pressure of the liquid of this output, which is itself assumed to be at atmospheric pressure. The molar mass of the gas blanket is:

$$M_E = \pi(t_E) M_V + (1 - \pi(t_E)) M_A$$

M_V and M_A are the molar masses of gas and air:

$$\rho_E = \frac{M_E}{22.4} \times \frac{273}{273 + t_E}$$

We know that there is a constant relationship between two successive nominal diameters.

$$0.8 = \frac{40}{50} = \frac{50}{63} = \frac{63}{80} = \frac{80}{100}$$

We can select ratio r accordingly, so that:

$$r = \frac{A_R}{A_E} = 0.8^2 = 0.64 < 1 \qquad [10.4]$$

A_A is the surface of the conical frustum between the vent and the inverted funnel. This surface is the difference between the surfaces of the following cones:

– a cone with cone distance ℓ;

– a cone with cone distance $R_E/\sin 50°$.

Finally:

$$A_A = \pi\left(l^2 \sin 50° - \frac{R_E^2}{\sin 50°}\right)$$

From A_E, we can deduce ℓ. We then trace a straight line at an angle of 50° to vertical from the edge of the vent. We place OC = ℓ along this line, which provides the position of the inverted funnel.

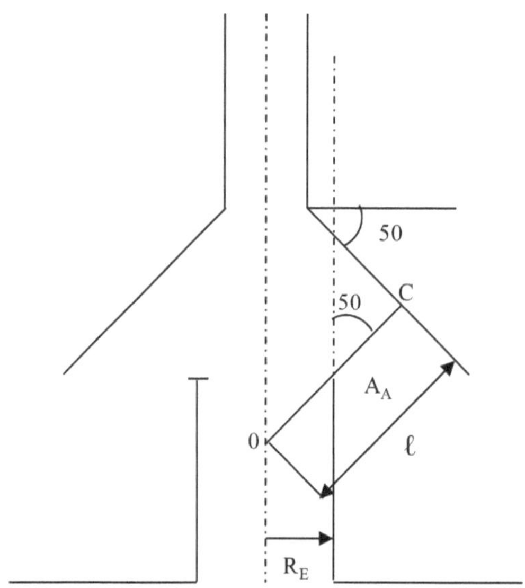

Figure 10.2. *Position of inverted funnel*

The value of relationship x defines W_A and A_A by equations [10.1] and [10.2].

The same value of x can be attributed to all vents. For this, we write the *approximate* result:

$$Q_V = Q_F + (1+x)Q_{ET} \quad [10.5]$$

Q_V: nominal fan flowrate ($m^3.s^{-1}$)

Q_F: set flow (for example, ventilation in a shop) ($m^3.s^{-1}$)

Q_{ET}: total flow emitted by the primary lines of vent ($m^3.s^{-1}$).

This relationship gives the value of x for the whole installation. We then use this value to determine each vent.

We should note that, even if flow W_E emitted by an output is cancelled, air is still sucked by the fan without the output falling into negative pressure at all.

In this situation, we see in relationship 10.3 that x is still very significant while τ becomes:

$$\tau \rightarrow \frac{r\rho_R}{\rho_A} \# r < 1 \quad [10.6]$$

In order for the value of τ provided by equation [10.1] to vary somewhat while x increases indefinitely, the values of Q must be such that:

$$x = \frac{Q_V - Q_F}{Q_{ET}} - 1 \quad \text{or as large as possible}$$

In other words, nominal fan flowrate Q_V must be significant in order to always *take in* enough air. Thus, the machine is in no danger of surging. Indeed, one criterion of non-surging is for the flowrate crossing the vent to always be greater by half of its nominal flowrate.

NOTE.–

The velocity V_G of gases in a pipe must always be about 10 $m.s^{-1}$ and, more precisely:

$$8\ m.s^{-1} < V_G < 14\ m.s^{-1}$$

The pressure drop in a pipe is:

$$\Delta P_t = f\frac{\rho_G V_G^2}{2}\frac{L}{D} \quad \text{with} \quad f = 0.03$$

The driving pressure ΔP_V of the fan must be:

$$\Delta P_V \geq 1,2\ \Delta P_t$$

NOTE.–

Let us use a constant gas flow.

The volume flowrate is maximum during warm weather.

$$t = 40°C \text{ inside} \quad \text{and} \quad t = 50°C \text{ outside}$$

Pressure drop in a pipe is minimum during cold weather.

$$t = +5°C \text{ inside} \quad \text{and} \quad t = -18°C \text{ outside}$$

These ranges apply to Europe.

NOTE.–

The arrival of air through cross-section A_A occurs in an annular convergent. As the velocity increases, the pressure must decrease. We have not taken this into account in the present summary. Curious readers may refer to Chapter 1, concerning vapor ejectors.

10.4. Nature of gas–liquid flows

10.4.1. *Horizontal flow or slightly sloping flow at angle α*

Taitel and Dukler [TAI 76] distinguish five types of flow:

– intermittent;

– stratified with waves;

– stratified smooth;

– dispersed bubbles (small);

– annular.

These authors give the relationships defining the passage from one type to another:

1) between stratified (smooth or with waves) and intermittent (or annular);

2) between intermittent (or with dispersed bubbles) and annular;

3) between stratified smooth and stratified with waves;

4) between intermittent and with dispersed bubbles.

Figures 5, 6 and 7 of Taitel and Dukler [TAI 76] allow us to situate the different domains.

10.4.2. *Vertical rising flow*

Taitel *et al.* [TAI 80] distinguish four main types of flow:

– with small or large bubbles;

– with pockets of gas interposed between liquid pistons (or plugs);

– destructured;

– annular.

In Figure 2 of their work, the authors present the boundaries between these domains:

1) between large bubbles and pockets;

2) between large and small bubbles;

3) between small bubbles and pockets (or destructured flow);

4) between pockets and destructured flow;

5) between destructured flow and annular flow.

10.4.3. *Vertical descending flow*

Barnea *et al.* [BAR 82a] distinguish three types of flow:

– with dispersed (large or small) bubbles;

– with pockets (separated by liquid pistons);

– annular.

These authors give the transitions between bubbles and pockets (of gas) and between pockets and annular flow.

10.4.4. *Sloping ascending or descending flow from horizontal to vertical*

Descending flow was addressed by Barnea *et al.* [BAR 82b] and inclined flow by Barnea *et al.* [BAR 85].

These authors provide a series of figures representing the progressive transition between horizontal and vertical flows.

NOTE.–

Wyckoff and Botset [WYC 36] studied the flow of a gas–liquid mixture in non-consolidated sand. This can be useful for the drying of porous bodies.

10.5. Pressure drop in gas–liquid flows

10.5.1. *The pipe is strictly horizontal*

Lockart and Martinelli [LOC 49] addressed the issue of liquid and gas hold-up together with that of pressure drop:

First, we calculate the physical property parameter:

$$\Psi = \left[\frac{\rho_G}{\rho_L}\right]^{0,5} \left[\frac{\mu_L}{\mu_G}\right]^{0,1}$$

The inverse I_x of Lockart and Martinelli's coefficient X is provided by:

$$I_x = \left[\frac{x}{1-x}\right]^{0,9} / \Psi$$

x: gas mass fraction in the mixture.

We then write out:

$$\phi^2 = 1 + 21\, I_x + I_x^2$$

Pressure drop ΔP is deduced from value ΔP_L, which is obtained if all the flow was in a liquid state:

$$\Delta P = \phi^2 \Delta P_L$$

Liquid hold-up is then provided by:

$$R_L = \phi^{-1}$$

And gas hold-up is deduced:

$$R_G = 1 - R_L$$

10.5.2. *The pipe is inclined, ascending or descending*

This situation occurs with the flow of oil in an oil pipeline. Effectively, natural gas is always inevitably mixed with oil. The slope of oil pipelines cannot exceed 20° either ascending or descending.

Singh and Griffith [SIN 70] showed that there is an optimal diameter for oil pipelines, that is, a diameter at which the pressure drop is minimum.

The authors give a very clear example for the pressure drop calculation.

10.5.3. *Hold-ups and pressure drop on ascent*

Hughmark and Pressburg [HUG 61] provide us with a means of calculating the hold-ups and pressure drop on ascent from a strictly vertical well.

Beggs and Brill [BEG 73] studied the hold-ups and pressure drop for an ascent of variable angle from 0 to 90 degrees to vertical.

NOTE.–

Dukler *et al.* [DUK 64] had attempted to calculate the pressure drop by using adimensional clusters. However, the authors admit that this procedure is not precise.

10.6. Critical biphasic flow

10.6.1. *Hypothesis*

This type of flow, in which the mixture velocity reaches the speed of sound, necessarily corresponds to high speeds, that is, the corresponding regime must be:

– either the annular regime with drops entrainment;

– or the bubble dispersion regime.

In both cases, we can consider that:

– the mixture is homogenous;

– the shifting speed between gas and liquid is negligible (isokinetic hypothesis);

– the gas and liquid are in thermal equilibrium, that is, at the same temperature (effectively, the discontinuous phase is considered to be finely divided).

10.6.2. *Speed of sound*

1) Liquids:

$$C_L = \sqrt{\frac{1}{\rho_L \chi}}$$

χ is the compressibility coefficient:

$$\chi = \frac{1}{\rho}\frac{d\rho}{dP} \qquad \left(Pa^{-1}\right)$$

2) Gases:

$$C_G = \sqrt{\gamma \frac{RT}{M}}$$

γ: ratio of specific heats (C_p/C_v)

R: perfect gas constant (8.314 $J.kmol^{-1}.K^{-1}$)

T: absolute temperature (K)

M: mean molar mass of gas ($kg.kmol^{-1}$)

3) Biphasic gas–liquid mixture:

Where x is the gas fraction in mass, the density ρ of the mixture is:

$$\frac{1}{\rho} = \frac{x}{\rho_G} + \frac{(1-x)}{\rho_L}$$

Hence:

$$\frac{d\rho}{\rho^2} = x\frac{d\rho_G}{\rho_G^2} + (1-x)\frac{d\rho_L}{\rho_L^2}$$

And:

$$\frac{d\rho}{dP} = x\left[\frac{\rho}{\rho_G}\right]^2 \frac{d\rho_G}{dP} + (1-x)\left[\frac{\rho}{\rho_L}\right]^2 \frac{d\rho_L}{dP}$$

But, by definition of speeds C_G and C_L:

$$\frac{1}{C_G^2} = \frac{d\rho_G}{dP} \quad \text{and} \quad \frac{1}{C_L^2} = \frac{d\rho_L}{dP}$$

Hence:

$$\frac{1}{C_{BP}^2} = x\left[\frac{\rho}{\rho_G C_G}\right]^2 + (1-x)\left[\frac{\rho}{\rho_L C_L}\right]^2$$

10.6.3. *Critical flow without friction*

According to our hypotheses, we can consider the biphasic mixture as homogenous and apply the impulse equation:

$$\rho u du + dP = 0$$

On the other hand supposing that the flow density at mass G is *maximum* at a given location:

$$G = \rho u: \quad \text{hence} \;\; \rho du + u d\rho = 0$$

Or rather:

$$\rho u du = -u^2 d\rho$$

And, applying this in the impulse equation (momentum equation):

$$u^2 = \frac{dP}{d\rho} = C^2$$

Thus, the mixture velocity is equal to the speed of sound.

10.6.4. *Flow equations*

These are:

1) Continuity (conservation of mass):

$$\frac{d\rho}{\rho} + \frac{du}{u} + \frac{dA}{A} = 0$$

2) Impulse:

$$\frac{dP}{\rho} + u du + u^2 4C_j \frac{dx}{2D} = 0$$

But:

$$\frac{dP}{\rho} = \frac{\chi}{\rho^2} d\rho$$

$$\frac{\chi}{\rho}\frac{d\rho}{\rho} + u^2 \frac{du}{u} + u^2 4C_f \frac{dx}{2D} = 0$$

3) Energy:

$$C_p dT + u^2 \frac{du}{u} = 0$$

With:

$$C_p = \frac{x\gamma R}{\gamma - 1} + (1 - x) C_L = \text{const.}$$

4) State:

$$1/\rho = \frac{x}{\rho_G} + \frac{1-x}{\rho_L} = \frac{x\,RT}{MP} + \frac{(1-x)}{\rho_{L0}} e^{\chi(P_0 - P)}$$

We thus have four equations for four unknowns: u, T, P and ρ.

We will not stress on the resolution of this system of equations.

NOTE (Limit flow).–

For a gas considered ideal, whose temperature is near constant, we can write:

$$\frac{dP}{P} = \frac{d\rho_G}{\rho_G}$$

The acceleration term ΔP_3 of Duckler's equation [DUC 64] can then be written as:

$$\Delta P_3 = \left[\left[\frac{-G_L^2}{R_L^2 \rho_L} + \frac{G_G^2}{R_G^3 \rho_G}\right]\frac{\partial R_L}{\partial P} - \frac{G_G^2}{R_G \rho_G P}\right]\Delta P = \alpha \Delta P_{global}$$

The value inside the global brackets is adimensional.

On the other hand, we know that:

$$\Delta P_{global} = \Delta P_{1_{friction}} + \Delta P_{2_{drop}} + \Delta P_{3_{acceleration}}$$

Hence:

$$\Delta P_{global} = \frac{\Delta P_1 + \Delta P_2}{1-\alpha}$$

Consequently, if coefficient α is close to 1, the global pressure drop increases rapidly. But before α can be equal to 1, we approach the speed of sound, which is even more the case since the speed of sound in a gas–liquid mixture is as low as 30 m.s^{-1}.

NOTE (Vapor–liquid mix).–

Condensation or vaporization can occur in the mixture and vapor mass fraction x is not invariable. On the other hand, the pressure and temperature are not independent but connected by the vapor saturation pressure expression:

$$P = \pi(T)$$

Fraction x is obtained from the energy equation that is written as:

$$d\left[xH(T) + (1-x)h(T)\right] + udu = 0$$

$$\left[H(T) - h(T)\right]dx + \left[x\frac{\partial H(T)}{\partial T} + (1-x)\frac{\partial h(T)}{\partial T}\right]\frac{dT}{dP}dP$$
$$+\left[x\frac{\partial H(T)}{\partial P} + (1-x)\frac{\partial h(T)}{\partial P}\right]dP + udu = 0$$

With:

$$\frac{dT}{dP} = 1/\left[d\pi(T)/dT\right]$$

Hence, dx according to dP.

Appendix

Characteristics of Various Gases

Gas	Molar mass	γ (15°C)	P_{cri}	T_{cri}
Acetylene	26	1.24	62.4	172
Air	29	1.40	37.7	133
Ethyl alcohol	46	1.13	63.9	517
Methyl alcohol	32	1.20	79.8	513
Ammoniac	17	1.31	114.2	406
Argon	40	1.66	48.6	151
Nitrogen	28	1.40	33.9	127
Benzene	78	1.12	49.2	563
Isobutane	58	1.10	36.5	408
n-Butane	58	1.09	38	425
Isobutylene	56	1.10	40	418
Butylene	56	1.11	40.2	420
Chlorine	71	1.36	77.1	417
Ethyl chloride	64.5	1.19	52.7	460
Methyl chloride	50.5	1.20	66.7	417
n-Decane	142	1.03	22.1	619
Carbon dioxide	44	1.30	74	304
Sulfur dioxide	64	1.24	78.7	430
Ethane	30	1.19	48.8	305
Ethylene	28	1.24	51.2	283
Water gas	19.5	1.35	31.3	130
Catalytic cracker gas	28.8	1.20	46.5	286

Coke oven gas	10.7	1.35	29.1	109
Blast furnace gas	29.6	1.39	–	–
Natural gas	18.8	1.27	46.5	210
Helium	4	1.66	2.3	5
n-Heptane	100	1.05	27.4	540
n-Hexane	86	1.06	30.3	508
Hydrogen	2	1.41	13	33
Methane	16	1.31	46.4	191
Carbon monoxide	28	1.40	35.2	134
n-Nonane	128	1.04	23.8	596
n-Octane	114	1.05	24.9	569
Oxygen	32	1.40	50.3	154
Isopentane	72	1.08	33.3	461
n-Pentane	72	1.07	33.7	470
Propane	44	1.13	42.5	370
Propylene	42	1.15	46	365
Hydrogen sulfide	34	1.32	90	374
Steam	18	1.33	221.2	648

Bibliography

[ANO 77] ANONYMOUS, "Toward perfect screw design", *Modern Plastics International*, pp. 59–59, November 1977.

[BAR 82a] BARNEA D., SHOHAM O., TAITEL Y., "Flow pattern transition for vertical downward two phase flow", *Chemical Engineering Science*, vol. 37, pp. 741–744, 1982.

[BAR 82b] BARNEA D., SHOHAM O., TAITEL Y., "Flow pattern transition for downward inclined two phase flow: horizontal to vertical", *Chemical Engineering Science*, vol. 37, pp. 735–740, 1982.

[BAR 85] BARNEA D., SHOHAM O., TAITEL Y. *et al.*, "Gas–liquid flow in inclined tubes: flow pattern transitions for upward flow", *Chemical Engineering Science*, vol. 40, pp. 131–136, 1985.

[BEG 73] BEGGS H.D., BRILL J.P., "A study of two-phase flow in inclined pipes", *Journal of Petroleum Technology*, vol. 25, pp. 601–617, 1973.

[BOO 63] BOOY M.L., "Influence of channel curvature on flow, pressure distribution and power requirements of screw pumps and melt extruders", *Society of Plastic Engineers Transactions*, vol. 3, no. 3, pp. 176–185, 1963.

[BOU 68] BOUSSINESQ M.J., "Mémoire sur l'influence des frottements dans les mouvements réguliers des fluides", *Journal de Mathématiques Pures et Appliquées*, vol. 13, pp. 377–424, 1868.

[BRO 46] BROWN G.P., DI NARDO A., CHENG G.K. *et al.*, "The flow of gases in pipes at low pressures", *Journal of Applied Physics*, vol. 17, pp. 803–813, 1946.

[BRO 51] BROMLEY L.A., WILKE C.R., "Viscosity behaviour of gases", *Industrial and Engineering Chemistry*, vol. 43, pp. 1941–1648, 1951.

[BRO 72] BROYER E., TADMOR Z., "Solids conveying in screw extruders. Part I: a modified isothermal model", *Polymer Engineering and Science*, vol. 12, no. 1, pp. 12–24, 1972.

[BRO 93] BROWN H.R., HANSON R.L., "Venting dust explosion", *Chemical and Metallurgical Engineering*, vol. 40, p. 116, 1993.

[BRU 68] BRUN E.A., MARTINOT-LAGARDE A., MATHIEU J., *Mécanique des Fluides*, vol. II, Dunod, 1968.

[BRU 70] BRUN E.A., MARTINOT-LAGARDE A., MATHIEU J., *Mécanique des Fluides*, vol. III, Dunod, 1970.

[CAR 53] CARLEY J.F., MALLOUK R.S., MC KELVEY J.M., "Simplified flow theory for screw extruders", *Industrial and Engineering Chemistry*, vol. 45, no. 9, pp. 974–978, 1953.

[CHU 63] CHUNG C.I., "A new theory for single-screw extrusion. Part 1", *Modern Plastics*, vol. 45, no. 13, pp. 178–198, 1963.

[CHU 68] CHUNG C.I., "A new theory for single-screw extrusion. Part 2", *Modern Plastics*, vol. 45, no. 16, pp. 110–115, 1968.

[CHU 70] CHUNG C.I., "New ideas about solids conveying in screw extruders", *Society of Plastic Engineers Journal*, vol. 26, no. 5, pp. 32–44, 1970.

[CHU 74] CHUNG C.I., NICHOLS R.J., KRUDER G.A., "Screw horsepower dependence on screw speed and size. Theory and experiment", *Polymer Engineering and Science*, vol. 14, no. 1, pp. 28–37, 1974.

[CHU 75] CHUNG C.I., "Maximum pressure developed by solid conveying force in screw extruders", *Polymer Engineering and Science*, vol. 15, no. 1, pp. 29–34, 1975.

[CHU 77] CHUNG C.I., "A guide to better extruder screw design", *Plastics Engineering*, pp. 34–37, February 1977.

[DAR 81] DARBY R., MELSON J., "How to predict the friction factor for flow of Bingham plastics", *Chemical Engineering*, pp. 59–61, 28 December 1981.

[DE 59] DE FRATE L.A., HOERL A.E., "Optimum design of ejectors using digital computers", *Chemical Engineering Progress Symposium Series*, vol. 55, no. 21, pp. 43–51, 1959.

[DON 71] DONOVAN R.C., THOMAS D.E., LEVERSON L.D., "An experimental study of plastificating in a reciprocating-screw injection molding machine", *Polymer Engineering and Science*, vol. 11, no. 5, pp. 353–360, 1971.

[DUC 64] DUCKLER A.E., WICKS M., CLEVELAND R.G., "Frictional pressure drop in two-phase flow: B. An approach through similarity analysis", *American Institute of Chemical Engineers Journal*, vol. 10, pp. 44–51, 1964.

[ECC 58] ECCHER S., VALENTINOTTI A., "Experimental determination of velocity profiles in an extruder screw", *Industrial and Engineering Chemistry*, vol. 50, no. 5, pp. 829–836, 1958.

[EDM 61] EDMISTER W.C., *Applied Hydrocarbon Thermodynamics*, Gulf Publishing Company, 1961.

[GRE 01] GRESH M.T., *Compressor Performance Aerodynamics for the User*, 2nd ed., Butterworth-Heinemann, 2001.

[GRI 62] GRIFFITH R.M., "Fully developed flow in screw extruders", *Industrial & Engineering Chemistry Fundamentals*, vol. 1, pp. 180–187, 1962.

[GÜL 89] GÜLICH J.F., RÖSCH A., "Cavitation erosion in centrifugal pumps", *Chemical Engineering Progress*, vol. 85, pp. 68–73, November 1989.

[HUG 61] HUGHMARK G.A., PRESSBURG B.S., "Holdup and pressure drop with gas–liquid flow in a vertical pipe", *American Institute of Chemical Engineers Journal*, vol. 7, no. 4, p. 677, 1961.

[JEA 54] JEAGER K., *Hydraulique Technique*, Dunod, 1954.

[JEP 53] JEPSON C.H., "Future extrusion studies", *Industrial and Engineering Chemistry*, vol. 45, no. 5, pp. 992–993, 1953.

[KNU 09] KNUDSEN M., "Die gesetze der Molekularströmung und der inneren Reibungsströmung der Gase durch Röhren", *Annalen der Physik (Leipzig)*, vol. 28, pp. 75–130, 1909.

[MAD 59] MADDOCK B.H., "A visual analysis of flow and mixing in extruder screws", *Society of Plastic Engineers Journal*, vol. 15, pp. 383–389, 1959.

[MAD 64] MADDOCK B.H., "Measurement and analysis of extruder stability", *Society of Plastic Engineers Journal*, vol. 20, pp. 1277–1283, 1964.

[MAD 72] MADDOCK B.H., SMITH D.J., "Extruder design by computer printout", *Society of Plastic Engineers Journal*, vol. 28, no. 5, pp. 12–17, 1972.

[MAD 74a] MADDOCK B.H., "Extruder scale-up by computer", *Polymer Engineering and Science*, vol. 14, no. 12, pp. 853–858, 1974.

[MAD 74b] MADDOCK B.H., "Barrier screw design holds no obstacles for computers", *Plastics Engineering*, pp. 28–33, January 1974.

[MET 65] METCALF J.R., "The mechanics of the screw feeder", *Proceedings of the Institution of Mechanical Engineers*, vol. 180, pp. 131–140, 1965.

[MIL 23] MILLIKAN R.A., "Coefficients of slip in gases and the law of reflection of molecules from the surfaces of solids and liquids", *The Physical Review*, vol. 21, no. 3, p. 217, 1923.

[MOH 61] MOHR W.D., CLAPP J.B., STARR F.C., "Flow patterns in a non-Newtonian fluid in a single-screw extruder", *Society of Plastic Engineers Transactions*, vol. 1, no. 3, pp. 113–120, 1961.

[MOO 90] MOODY F.J., *Introduction to Unsteady Thermofluid Mechanics*, John Wiley & Sons, 1990.

[MUT 01] MUTAFTSCHIEV B., *The Atomistic Nature of Crystal Growth*, Springer, 2001.

[PIN 70] PINTO G., TADMOR Z., "Mixing and residence time distribution in melt screw extruders", *Polymer Engineering and Science*, vol. 10, no. 5, pp. 279–288, 1970.

[SCO 62] SCOTT D.S., DULLIEN F.A.L., "The flow of rarefied gases", *American Institute of Chemical Engineers Journal*, vol. 8, no. 3, p. 293, 1962.

[SHE 75] SHERIDAN L.A., "The practical side of mixing in extruders", *Chemical Engineering Progress*, vol. 71, no. 2, pp. 83–84, 1975.

[SHI 74] SHIMIZU J., "Screw type continuous mixing mechanisms and their trends", *Japan Plastic Age*, pp. 30–38, August 1974.

[SIN 70] SINGH G., GRIFFITH P., "Determination of the pressure drop optimum pipe size for a two-phase slug flow in an inclined pipe", *Journal of Engineering for Industry*, vol. 92, pp. 717–726, 1970.

[SQU 58] SQUIRES P.H., "Screw extruder pumping efficiency as influenced by relative channel depth and curvature", *Society of Plastic Engineers Journal*, vol. 14, no. 5, pp. 24–30, 1958.

[STR 61] STREET L.F., "Plastifying extrusion", *International Plastics Engineering*, pp. 289–295, July 1961.

[TAD 66a] TADMOR Z., "Fundamentals of plasticating extrusion", *Polymer Engineering and Science*, vol. 6, no. 3, pp. 185–190, 1966.

[TAD 66b] TADMOR Z., "Non-Newtonian tangential flow in cylindrical annuli", *Polymer Engineering and Science*, vol. 6, no. 3, pp. 203–212, 1966.

[TAD 67] TADMOR Z., DUVDEVANI I.J., KLEIN I., "Melting in plasticating extruders. Theory and experiments. Part I", *Polymer Engineering and Science*, vol. 7, no. 3, pp. 198–217, 1967.

[TAD 69] TADMOR Z., KLEIN I., "The effect of design and operating conditions on melting in plasticating extruders", *Polymer Engineering*, vol. 9, no. 1, pp. 1–10, 1969.

[TAD 71a] TADMOR Z., BROYER E., KLEIN L., "Solids conveying and melting in single screw extruders. Part I", *Proceedings of the 29th Annual Technical Conference (ANTEC)*, Washington, DC, vol. 12, no. 1, pp. 136–140, 10–13 May 1971.

[TAD 71b] TADMOR Z., KATSIR L., KLEIN I., "Solids conveying and melting in single screw extruders. Part II. Beginning of melting", *Proceedings of the 29th Annual Technical Conference (ANTEC)*, Washington, DC, pp. 141–147, 10–13 May 1971.

[TAI 76] TAITEL Y., DUKLER A.E., "A model for predicting flow regime transitions in horizontal and near horizontal gas–liquid flow", *American Institute of Chemical Engineers Journal*, vol. 22, no. 1, p. 47, 1976.

[TAI 80] TAITEL Y., BORNEA D., DUKLER A.E., "Modelling flow pattern transitions for steady upward gas–liquid flow in vertical tubes", *American Institute of Chemical Engineers Journal*, vol. 26, no. 3, p. 345, 1980.

[TEM 80] TEMPERLEY H.N.V., TREVENA D.H., *Les Liquides et Leurs Propriétés*, Technique et Documentation, 1980.

[WHE 62] WHEELER N.C., "Junior common extrusion problems", *Wire and Wire Products*, vol. 37, pp. 1743–1747, 1962.

[WIL 50] WILKE C.R., "A viscosity equation for gas mixtures", *The Journal of Chemical Physics*, vol. 18, no. 4, p. 517, 1950.

[WYC 36] WYCKOFF R.D., BOTSET H.G., "The flow of gas–liquid mixtures through unconsolidated sands", *Journal of Applied Physics*, vol. 7, pp. 325–345, 1936.

[WYM 75] WYMAN C., "Theoretical model for intermeshing twin screw extruders: axial velocity profile for shallow channels", *Polymer Engineering and Science*, vol. 15, no. 8, pp. 606–611, 1975.

Index

O, P, R

S, T

V, Y

FSC
www.fsc.org
MIX
Papier | Fördert gute Waldnutzung
FSC® C083411

Kolibri 360 GmbH
Ferdinand-Jühlke-Straße 7
99095 Erfurt, Deutschland
produktsicherheit@kolibri360.de